ISBN 978-0-9953712-0-0

www.eliminateautomateoffshore.com

# ELIMINATE
# AUTOMATE
# OFFSHORE

## Why Our Careers
## Are Facing Extinction

Rob Gaunt

# Preface

If you are too young to remember or perhaps the parent of someone who is too young to remember a time when the only two types of telephones weren't Android or Apple but those connected to the wall in your house and those in phone booths, then this book is for you.

Unless you're independently wealthy, at some point you'll find yourself having to choose a career path. It would be a good idea to select one that is likely to still be around in 10, 15 and 20 years' time because, as this book explains, most won't be.

If you don't believe me, next time you take a flight ask one of the cabin crew for the name of the navigator......

The key to selecting a career with longevity can be summarised as follows:

There is only a future for careers in expensive economies where the value or quality of local knowledge or presence outweighs the cost advantage of every other method of performing the task.

This book is not a how-to guide. It is a skip through the recent history of work with a look forward to the future and is designed to be thought-provoking.

No one can predict the future. Nevertheless, we should all be preparing for it. The following pages might just help.

# Contents

# Introduction

As parents, we want to give our children the benefit of our experiences and the insights we've learned over our lives. But what if those experiences bear little relation to the world they will grow up in?

If, like me, you suspect this might be the case, read on.

For the last decade and a half I have worked as an independent management consultant advising client organisations on how to cut costs. In large organisations, one of the biggest costs is the workforce.

Every organisation that doesn't have access to an unlimited source of money (i.e. taxes) must constantly strive to employ fewer people and find ways to ensure the remaining staff are more productive.

This is what I do. In other words, I'm a corporate axe-man, "the toe-cutter", the most disliked person in the company. You may not approve of or like what I do but that doesn't mean it won't happen.

I don't take pleasure in firing people and I try to do it in the most professional and sympathetic way possible, but if the role is no longer economically viable, it is going to be made redundant by me or someone else.

My mortgage is paid by me, not someone else, giving the bad news to people. So that's what I do.

Like many people, I've come to the conclusion over recent years that the world of work is going to be very different to the one my generation faced when we left school and college. What was considered an honourable, prestigious or "safe" profession may be unrecognisable or might not even exist in the new world of work. In the meantime, schools and universities are still teaching a curriculum that, for the most part, a pupil of 30 years ago would have recognised and understood.

Despite this, jobs are disappearing in front of our eyes. We can find examples everywhere we look;

• Unskilled labour - my local council waste collection vehicle now uses a hydraulic arm which lifts and tips the bin in to the truck. Five years ago that used to be 4 people per vehicle to move the bin to the back of the truck. Fifteen years ago, a team of perhaps eight to 10 men used to physically lift the bins from the roadside to the back of the truck. Now it's an additional task for the driver, and one for which he's not paid significantly more to perform (certainly not the salaries of the nine missing team members).

• Semi-skilled labour – the newspaper you no longer read in paper form doesn't require a typesetter. The relatively few printed copies sold require barely any humans in the production process. These were jobs which used to offer apprenticeships and were considered safe career options. So much so that many newspaper print rooms were "closed shops", that is, the union determined who could and couldn't be hired in to new vacancies and the best jobs were bequeathed from father to son.

• Highly-skilled labour – There is one less job on board modern aircraft from their equivalent in the 1980s. The navigator used to have a dedicated seat in the cockpit of a commercial airplane and was paid well for the skills and experience required to keep the pilot flying on the correct course. Nobody has the job title "navigator" in modern airlines because the function is now performed by a slight glance at the GPS display by the pilot or co-pilot.

My work provides a unique insight into many of the factors driving the changes to the future workforce.

When I think about what my children might do for a living in the future, I think about the jobs I currently can't eliminate, automate or send offshore.

What is special about these jobs?

Is it just a matter of time before these are no longer required, can be undertaken by a machine or can be performed to an acceptable level of quality in a lower cost economy?

If not, why not?

Perhaps these are the jobs I should encourage my children to aim for.

# Chapter 1
# A new industrial revolution is building a new economy
## A history of humans in three revolutions

The earliest evidence suggests Homo Sapiens evolved around 160,000 years ago. Assuming that a generation has a duration of between 20 and 30 years, we can estimate that there have been around 5,500 generations of our species so far.

The first 5,000 of these survived in an exclusively hunter/gatherer existence. Life and death occurred along roughly the same pattern from father to son, mother to daughter for much of human history; following the herds, picking fruits and berries and living at the mercy of the seasons and the climate.

Ignoring climatic and dietary differences, this existence was common across every continent inhabited by humans.

Then, around 11,500 years ago, modern agriculture was invented and, for the last 500 or so generations, our lives changed to follow a different pattern; waking at dawn, tending the fire, tilling the crops, feeding the animals, perhaps hand-spinning yarn and weaving thread for clothes and settling down at sunset until the following day.

With few exceptions, children could expect to have the same pattern of life as their parents across these "modern" generations.

Between 1760 and 1840, new discoveries and inventions permanently changed how humans would live.

The previous certainty that what you did to survive or earn a living would be very similar to what your children would do was replaced with the inverse statement as an almost certainty; the work routines of the next generation were practically guaranteed to be very different to their parents' generation.

People living at the start of what we now call The Industrial Revolution, roughly 10 generations ago, were only the second generation of humans in history to experience this.

We are now the third generation in human history to experience such a massive upheaval to our pattern of life. To emphasise this point, we have become the first generation where urban dwellers outnumber rural dwellers; this switch occurred around 2007.[1]

The changes we are currently living through will have as much impact on human existence as the discovery of techniques to cultivate grain or the invention of the spinning jenny.

Interestingly, despite our acceptance of the many benefits of these changes, many of us are blithely unaware or uninterested at the enormity of the upheaval to human existence and the implications to our descendants' lives.

Look again at the rate of acceleration of these revolutions in human existence. The number of generations living in an age of relative certainty are reducing exponentially from the evolution of the species to the present day; 5,000 generations, 500 generations, 10 generations, the current generation.

---

[1] *http://beta.sedac.ciesin.columbia.edu/data/collection/gpw-v3*

Taking as an example the manufacture of clothing, we can see how these human revolutions have impacted those involved in the process over time;

In the period between 160,000 and 11,500 years ago (i.e. the first 5,000 generations) humans used animal skins as coverings. Innovations then followed such as treatments to soften the leather to provide the wearer more freedom of movement and the invention of bone or wooden needles to stitch the furs together to make the seams more windproof, thus extending the temperatures and conditions in which the wearer could be active.

After the domestication of crops and animals for flesh and milk around 11,500 years ago, sheep began to be selected and bred for their fleeces. The process to hand-spin yarn and knit clothes remained broadly unchanged until the industrial revolution.

The Industrial Revolution in Britain 250 years ago saw inventions such as John Kay's "flying shuttle" and James Hargreaves' "spinning jenny" changing the methods of clothes' production forever. There then followed a rapid series of disruptive innovations and a centralisation of the workers into custom-built factories with new sources of power (Arkright's Mill, the first water-powered factory, for example).

Mill owners looking to improve their profits realised that key steps in the process of production could be performed faster, cheaper and with more certain quality outcomes if a machine did the work.

Around the same time, trade between the British colonies opened up new sources of raw material and markets to sell the finished products to. The British began to import cotton harvested from its colonies in India and America, process this to manufacture cloth which was then exported for sale back to the colonies.

In the early years of the 20th Century, the Indian independence movement raised consciousness around this loss of self-sufficiency, culminating in Mohandas Gandhi's Quit India campaign in the 1940s.

Gandhi used the symbol of the spinning wheel as a point of focus throughout the battle for independence and it remains on the Indian national flag to this day.

The direction of the trade in clothes has reversed in the intervening years, with the countries which emerged following the partition of British India (India, Pakistan and Bangladesh) switching to be net exporters of textiles to markets such as Britain.

Further mechanisation and automation of the process has also occurred. In the most modern textile factories hardly any humans are required compared to only a few decades ago with computer-controlled production lines delivering finished products which may have been designed overseas and the patterns remotely uploaded to the machinery.

It's obviously impractical to compare the production rate and cost in human hours worked of a garment from the dawn of human history to the present day, but if we could, it would be unsurprising to see the production rate rising and the human hours required dropping exponentially.

A fully-automated factory can produce thousands of Hello Kitty t-shirts in the time an ancestor from the Stone Age would have required to produce a single fur cape.

## We have lived through a revolution, did we even notice?

Following the Second World War, the global economy was one underpinned by restrictive labour practices, protective international trade tariffs and communication barriers.

The subsequent Cold War saw many markets closed or trade limited between countries on different sides of Churchill's "Iron Curtain".

Britain, for example, spent much of the period after the war until the early 1980s at the behest of a powerful trade union movement. Attempts to improve productivity and efficiency which would have resulted in a reduction in the workforce were frequently met with a strong reaction by union leaders, resulting in a febrile economy, reduced production and damaging strikes.

Recently-released government papers from "The Winter of Discontent" (1978/79)[2] describe the complete fear and paralysis affecting the governments of the day when negotiating with such formidable and, in many ways, unscrupulous negotiators.

Partly due to pressure from their internal labour markets or from protectionist political ideology, governments kept a tight control on the ability of other countries selling into their markets.

Import tariffs were used punitively against nations deemed to be unfriendly or threatening the longevity and salaries of the

---

2 *https://www.theguardian.com/uk/2008/dec/30/archives-callaghan-labour-memo-1978 but also many of the National Archives' Cabinet Papers such as CAB/129/204/25*

domestic workforces.

Even where explicit restrictions were legislated, international pressure was used to implement "voluntary restraint agreements" such as those forced on foreign imports of automobiles to the USA.

The price of oil has always been a significant factor in the free movement of international trade and, from the Arab Oil Embargo in 1973 (as a reaction to the West's support of Israel during the Yom Kippur War) through to the mid-1980s, this rose and remained high. The cost of transportation, particularly shipping, reflected this.

When businesses were able and could afford to transact across international borders, the communication options available to facilitate trade had significant limitations.

Telephone calls between countries were expensive and often unreliable compared to the present day.

Direct-dialling your counterpart in another country wasn't an option until relatively recently; rather, a call needed to be booked and scheduled with the operator.

Interactions which required written details or proof could only be transmitted either in the form of a physical letter or instantly (and therefore, expensively) over the telex network, but this had a speed limitation of around 65 words per minute so was not practical for the transmission of any complex information such as highly-technical specifications, detailed bills of materials or contracts for supply.

During the 1980s and 1990s, many of these restrictions to trade fell away as labour and import markets were de-restricted, the price of oil fell (due to the dilution of the power of the OPEC cartel following discoveries of new sources of supply such as those in the North Sea) and the Soviet economic system collapsed.

At the same time, economies such as India and, later, China began a series of economic reforms which started the process of transforming their economies away from being almost exclusively agrarian.

Not entirely coincidently, the 1980s and 1990s saw major technological advances; microchips and computing power rapidly moved from the exclusive domain of the military and large multinationals in industries such as finance and energy to something teenagers in garages in California could access and play around with.

New methods of connecting computers together in networks were invented. The internet was created and made available to a world-wide market.

As the cost of microchips reduced, more aspects of manufacturing, services and general life became subject to improvements in speed, efficiency and cost by the application of technology solutions.

It's difficult to find any examples of domestic, business or municipal activities which have remained unchanged by technology between the 1970s and the present day.

There is a clue here to one of the possible answers to the question posed in the introduction of this book.

One surprising source of the technological transformation came from the American military.

The decision by Ronald Reagan to allow civilian use of the

Defense Department's Global Positioning Satellite (GPS) network, following the shooting down of a Korean civilian aircraft in 1983, led to the ability of anyone on the planet to know their precise position to within a few metres from 1989 onwards.

Prior to this, to fix a position, the navigator on an aircraft flying in the 1970s used a handheld sextant (recognisable to Captain Cook and Lord Nelson) to take readings from the sun, moon and stars, then consult a book of complex tables and perform a series of calculations. President Reagan's announcement signalled the end of airline navigator as a career option to future generations.

Initially, the civilian use of the GPS network had a level of inaccuracy deliberately injected ("selective availability") which restricted users to an accuracy level to plus or minus 30 metres; this was adequate for the navigation of aircraft and ocean-going vessels.

The removal of selective availability in May 2000 gave the next step in granularity, enabling the current accuracy level of plus or minus one metre. This was the change which enabled everyday use, consigned the traditional paper road map to the garbage bin and enabled the launch of services such as Uber.

We are still inventing new ways to benefit from this new level of accuracy, from being able to call a taxi in a new city without being able to even read the street name to the not so useful invention of watching the progress of the moped delivering your pizza.

To a certain extent, the phenomenon of Silicon Valley also has its roots in the US military, via the funding provided by the

Defense Advanced Research Projects Agency (DARPA). This included the Internet and the Windows operating system amongst other advances.

But, before we become too hung up on the funding sources of the changes and begin travelling down the road of conspiracy theorists, it's worth considering these developments were likely to happen in one form or another because of all of the other contributing factors we've described above.

As Matt Ridley explains in his excellent book, *The Evolution of Everything*, most inventions were not a result of a moment of inspiration arriving in an exceptional brain but as a confluence of multiple discoveries and enabling technologies, suggesting another suitably open mind was also likely to have reached the same outcome around the same time.

Progress on determining the laws of natural selection, for example, was made by John Herschel in isolation to Charles Darwin. If Darwin had been shipwrecked on The Beagle, it is highly likely we would still have benefitted from a greater understanding of how evolution works.

Since the fall of the Berlin Wall and the switch of the Chinese economy to a more liberal version of Communism (one hesitates to call it capitalism; this is a long way from some Adam Smith utopia), major power shifts have occurred in the world economy.

A crude and anecdotal measure of these changes might be found on any summer day at a popular tourism destination such as the Spanish Steps in Rome; in 1980, my parents took our family on a sightseeing holiday to Italy. In addition to the socks-with-sandals-wearing English tourists and other nationalities, I clearly recall a coach-load of Japanese sightseers with expensive cameras around their necks.

In fact, through the 1970s and 1980s the camera-laden Japanese tourist became somewhat of a stereotype and the target of some fairly xenophobic humour in the British psyche.

Go to those same tourist destinations today and the Japanese numbers have been overtaken by South Koreans and Chinese. This unscientific observation reflects the shift in the world economy, in particular the "lost decades" the Japanese have experienced and the opening of the Chinese economy to the world, as the "plateauing" on the chart below demonstrates[3].

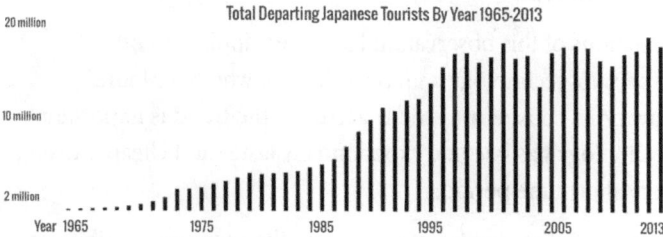

Total Departing Japanese Tourists By Year 1965-2013

Chinese outbound tourism was practically zero prior to 1985, at which point the Chinese government commenced a limited Visiting Friends and Relatives (VFR) visa for those living on the mainland to visit Hong Kong.

This freedom was expanded in 1991 to include organised visits to Singapore, Malaysia and Thailand. It was expanded again in 1997 when tourist visas were made freely available.

As a consequence, in 2012 83 million Chinese travelled abroad on holiday.

---

3 *http://www.tourism.jp/en/statistics/outbound/*

In the 1970s, the west began importing cheap imports from developing and low-cost economies such as Singapore, Hong Kong and Taiwan. Japanese manufacturing had a greater focus on quality electronics and the automotive market.

Latterly, China has taken over the vast majority of low-cost products and South Korea has taken over from Japan in the development and manufacture of high quality electronics. From a Western point of view, this process has had the effect of importing deflation.

The famous Moore's Law is an observation that the number of transistors in a dense integrated circuit doubles approximately every two years.

Variations of this observation have been applied to other aspects of technology and, regardless of whether Moore's Law or its cousins are strictly accurate, the trend is important; technology has become exponentially faster and cheaper over a very short time period.

This has led to significant changes in the uses and possibilities of computing power and telecommunications bandwidth.

International telecommunications, for example, has undergone an incredible increase in capacity and subsequent lowering of unit costs to the consumer.

The graph overleaf illustrates the capacity increase of channels over time (for simplicity, assume one channel is equivalent to the capacity required for a single telephone conversation)[4].

The capacity of transatlantic cables is not unique; at the same time similar levels of capacity were being introduced between the other continents and domestically in all major economies.

---

[4] https://en.wikipedia.org/wiki/Transatlantic_communications_cable

Transatlantic Telecommunications Capacity

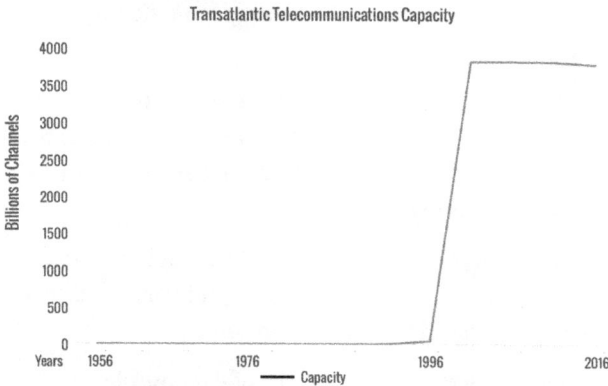

The explosion of telecommunications capacity from the mid-1990s has enabled wide-reaching trade and societal changes. One simply has to look at the founding dates of some of the new mega-companies which we all take for granted now to see the pattern;

**1994** Amazon

**1995** eBay

**1996** Google

**1999** Salesforce

**2002** LinkedIn

**2004** Facebook

None of these companies and services were possible without those additional telecommunications channels laid in the early and mid-1990s. Imagine what the Facebook "experience" would have been like in 1991 on a 9.6KB dial-up modem.

More recently, the telecommunications revolution has combined with computing and mobile phone technology to enable everybody to be online all the time from the small device in their pocket.

That small device has the equivalent computing power of the largest corporate data centre from 25 years ago; an iPhone 6 has more processing power than a Cray2 Supercomputer (the world's fastest computer until 1990).

We have the total combined documented human knowledge available at our fingertips and the ability to reference and share this with anyone else in the world at any time.

We have only just scratched the surface of the possible applications of this interconnectedness.

In parallel to the exponential increases to the technological and telecommunications capacity and lowering of the associated costs, came an increase in outsourcing.

In other words, the packaging up of discrete groups of business processes and passing these, often along with a transfer of the employment of the employees performing them to a third party organisation to undertake for a fee for service.

Outsourcing is simply a new word for something as old as the history of towns and villages; nobody thinks twice about buying shoes from a shoe shop or bread from a bakery, these are tasks we have chosen to delegate to specialists or to those who can offer us the benefits of the economy of scale.

The outsourcing of white collar processes isn't particularly new either; specialist accountancy services (e.g. giving rise to "the big four" of PWC, Deloitte, Ernst and Young and KPMG) and payroll processing services (sometime referred to as "bureaus") have been around for decades.

In the 1970s and 1980s, this accelerated with large corporations' IT departments being transferred over to other companies which would then deliver the services back to the original organisation for a service fee or, in the example of Deutsche Telekom's IT department, be spun off as a separate legal entity (T-Systems) to provide services as a commercial offering to the wider market in addition to the parent company.

Once the availability, reliability and cost of international telecommunications had improved significantly, the next logical import from low cost economies to western counties was the provision of services.

No longer were countries such as India, Mexico, the Philippines, China, and others limited to exporting competitively-manufactured physical products but they could now use the labour arbitrage advantage to undercut the cost of desk-bound jobs.

Over the last three decades, specialist companies have emerged to provide a range of business services from low cost economies to client organisations in the west.

IT support services were one of the first business processes to be provided remotely.

The expression "to be Bangalored" became slang in the USA for losing one's job due to offshoring (Bangalore is a large Indian city with one of the highest populations of IT services employees in the world).

As the business case and stability of these services were shown to be effective and adequate, further business processes were moved offshore. Back office accounting functions requiring high numbers of staff tended to be the next to be offshored and a wide range of other functions were soon to follow, as we will see in a case study later.

The preparations and activities required to successfully outsource and/or offshore a business function (or transition as it is called within the industry) have become standardised with most engagements following a fairly similar format;

- The services company undertakes an exercise to document and transfer knowledge of the processes from the incumbent staff,

- Telecommunications links are established from the remote office to the client's IT systems,

- Replacement staff are recruited and trained in the new processes,

- The services are cut-over to the new delivery team,

- The new operational management team commence running the service and produce reports on an agreed set of metrics to prove quality and volume is to the required levels (cost is determined by the contract).

In offshoring, the emotion and grief usually involved in being "let go" can be further exacerbated by the fact that you are likely to meet the team (or their colleagues) who will be taking your job from you and they will be expecting a quality handover of the what and how of your job in the period prior to the transfer.

In fact, the transition team's performance will be judged on how accurate and complete that handover is, so they will be particularly tenacious in their requests for information.

The changes to the work people have performed and the

process this chapter has described can be summarised by three verbs;

**Eliminate**

**Automate**

**Offshore**

Either as part of a conscious strategy or through a gradual process of market forces applying their relentless drive to find more efficient ways of delivering the same goods and services, work has been eliminated completely, partially or fully automated or delivered from a lower cost location.

This is a continuous cycle. The Chief Executive Officer of every publicly-listed or privately-owned company in every country of the world thinks about their costs in this way almost instinctively and if they don't, they'll soon discover their competitors do.

The history of business is littered with failed organisations which lost control of their cost base relative to their competition in this way.

The prime responsibility of a privately-owned business is to provide value to its shareholders, either by increasing the value of the stock or paying a dividend and preferably both. To achieve this, the business must make a profit and there are only two ways a business can do this; either increase the volume of sales or cut the cost base.

In a competitive market, the CEO must do both.

There's little point bemoaning the "evils" of capitalism either.

If the CEO is the head of a government department, municipal authority of Not for Profit (NFP) organisation, they are likely to have a legal responsibility to provide services to the taxpayers or beneficiaries at the best value possible.

Granted, the interpretation of this requirement and the enthusiasm with which it is implemented is usually far looser than the CEO of a company faced with having to explain higher costs relative to the competition to the shareholders at the Annual General Meeting, but the public sector has to drive out unnecessary costs, nonetheless.

In the next three chapters we will explore further the ways in which work is eliminated, automated and offshored.

# Chapter 2
# Eliminate
## "The unexamined life is not worth living" - Aristotle.

There are inefficiencies in the way we perform practically every activity in our work and domestic life.

For tasks which are performed rarely or only ever once, this is nothing to be particularly concerned about.

However, if the task is performed hundreds, thousands or perhaps hundreds of thousands of times, the wasted effort and time becomes a problem.

In business, this might be the difference between winning or losing business over your competitors due to higher prices, lower quality or delayed delivery times.

It's difficult to understand how anyone might find job satisfaction performing a task that is not required, but I have seen real examples of this many times in my career.

To understand how these situations might arise, consider the layout of your kitchen at home; the location of the various appliances, the sink, the refuse bin and the counter top surface you use for preparing food.

Now imagine the steps you need to take to prepare a toasted cheese sandwich; taking the bread from the fridge and placing two slices in a the sandwich toaster (is the toaster kept permanently on the counter top or in a cupboard?), taking the cheese from the fridge (hang on, is that the second visit to the fridge?), taking a plate from the cupboard (the same cupboard as the toaster?), taking a knife from the drawer to slice the cheese, putting the cheese between the two slices of bread, taking a fork from the drawer to use with the knife to eat the sandwich (is that the second time you've been to the drawer?) and then sitting at the table to eat.

Clearly, the "making a toasted cheese sandwich process" isn't the most efficient it could be.

We can console ourselves with the realisation the kitchen is also used in hundreds of different "processes" in addition to the cheese sandwich version, so is probably about as efficient as we need it to be, unless our entire diet exclusively consists of toasted cheese sandwiches.

However, if we were considering building a factory to make hundreds of thousands of our gourmet toasted cheese sandwiches, we would be well-advised to look at removing as many of the repeat steps as possible and running others in parallel.

Inefficiencies exist everywhere in the world of work.

Those organisations which can identify and address these inefficiencies quickly, and before they mount up to add significantly to the cost base, will have a competitive advantage.

The corollary of this is the example of British Leyland Motors, a vehicle manufacturer which staggered from threatened bankruptcy to repeated government bailouts throughout the 1970s and 1980s.

One of the more awful products produced at the Longbridge

factory was the Austin Allegro, a car which quickly gained a reputation for problems related to early rusting, partly due to the fact it was built in two separate sheds and the part-finished vehicles were moved, without cover and whatever the weather, between the two buildings!

At the same time, one of their overseas competitors, Japan's Toyota, were constantly reviewing their manufacturing processes using a methodology called "Kaizen" (Japanese for 'change for better') which had been incorporated at all levels of the organisation.

Kaizen uses a constant cycle of Plan/Do/Check/Act to ensure the way things were done previously does not have to be the way they have to be done in the future.

Participation in the Kaizen methodology was encouraged and achieved at all levels of the Toyota workforce.

It's hard to imagine the militant unionised workforce of the Longbridge plant in the late 1970s participating in Kaizen, especially when one the main outcomes of continuous improvement is the reduction of the number of staff required to achieve an equivalent outcome.

British Leyland is an extreme but not unique example of the type of thinking which can be summarised as "we do it this way because we've always done it this way".

People often choose not to question the "why" of what they do and, over time, the original reason for a step in a process, or indeed the entire process, is forgotten. In the words of the American author, Upton Sinclair, "'It is difficult to get a man to understand something, when his salary depends on his not understanding it".

Since the adoption of Kaizen across many industries, there have been further enhancements to the methodologies and techniques used to drive out inefficiencies in processes regardless of whether the end outcome is the manufacture of a new automobile with the highest quality and lowest price, the accurate calculation and timely payment of thousands of employees' weekly wages or the design and support of a new IT banking system.

These methodologies and tools include:

- Total Quality Management (TQM)
- ISO9000
- Lean
- Six Sigma
- Kaizen
- Takt Time Analysis
- Pareto Analysis
- Impact vs Effort Analysis

Regardless of which methodology and tool is favoured within a particular organisation, the outcome is that continuous improvement is no longer an afterthought to be sometimes undertaken depending on the diligence of a particular business leader.

The rate or effectiveness of improvements might still be inconsistent based on competency, motivation and capability of individual managers but most successful organisations have at least mandated these programmes of improvement and have benefitted from the outcomes they drive.

General Electric, for example, claimed it had saved $350 million from the implementation of the Six Sigma methodology over a three year period. This figure rose to $1 billion over the next few years, or the equivalent cost of nearly 13,000 employees on an average salary of $75,000.

## Every process can be improved, even waste collection

Let's return to the refuse truck example from the introduction. There have been major changes to the way domestic waste has been collected over the last 30 years.

Until the 1980s, domestic waste would be collected by a truck with a driver and a team of up to 10 collectors who manually carried cylindrical-shaped bins (purchased by the resident) on their backs to the rear of the truck where they would then empty the bins. Where I grew up in the UK, residents didn't even have to place the garbage bins by the side of the road on the designated collection day; the bin men actually entered the property to collect it (much to the amusement of our dog).

In countries such as the UK, the next innovation was the introduction of standardised box-shaped "wheelie" bins (provided by the local authority). These were wheeled to the rear of a new model of truck with a hydraulic lift to be automatically emptied. The number of men required to collect the bins was able to be reduced whilst still achieving the same levels of productivity. In addition, health and safety outcomes were improved as the work required less physical exertion and risk (from carrying potentially dangerous rubbish on their shoulders) and fewer men working on busy streets.

Today, many local authorities sub-contract to waste management companies who now use a collection truck with a hydraulic side arm operated by the driver (sitting on the opposite side of the cab to regular vehicles) – an Australian invention. The driver positions the truck such that the side-attached hydraulic arm is alongside the bins and then lifts and empties these into the middle of the truck without leaving the cab.

There is one additional member of the team, a "runner" who precedes the truck and driver by an hour or so to align the bins to be free of obstructions and away from parked cars. Again, in addition to the productivity efficiencies, the risk to the remaining workers has been massively-reduced.

By examining these very observable changes, it's obvious around eight or nine jobs per truck have gone forever, four of which were as a result of the not exactly inspiring invention of a bin with wheels combined with a mechanical hook and lever system on the truck.

The entirely driver-only vehicle then removed the remaining manual labourers' jobs, with the exception of probably the local authority's fittest employee who is still required to jog ahead to straighten the bins.

Some local authorities have even removed the requirement for this team member by communicating to the householders that bins not in the correct position won't be emptied.

What's interesting though, is the business case to completely change the fleet of trucks, pay redundancy liabilities to the fired bin collectors and provide a brand new bin to every resident must have at least provided a pay-back when balanced against the ongoing costs of paying the labourers' wages and injury claims.

The capital investment required to make each change to domestic waste collection described above is not insignificant; in Australia, a side-arm driver-operated truck costs approximately AUD $400k and the unit cost of a wheelie bin is around AUD $80.

Generally, the local authority waste collection contracts are awarded for a minimum of five years but longer contracts of up to 10 years are also awarded. The two main changes (switching to wheelie bins and then introducing driver-only collection trucks) delivered a positive return on investment for each local authority within this five to 10 year period including the capital investments made by the waste collection contractors.

## Containerisation

Prior to the late 1950s, nearly all goods were manhandled between various modes of transport. Trains would arrive at the port and teams of men with shovels, cranes and trolleys would move the various items between the ships in dock.

Men would stand in the holds of ships shovelling grain or coal into large buckets to be lifted by crane onto the dockside. In addition to the obvious health and safety risks, there were many disadvantages to this, not least of which the fact that the ship would often have to spend more time in dock than at sea, which, for an expensive asset, was highly inefficient.

The subsequent high rate of breakage, loss and theft also added to the cost overhead and had to be borne by the owner of the cargo and ultimately passed on to the consumer.

The cost to move goods across oceans would often be close to 25 per cent of the value of the cargo.

Although the idea of packing goods in to standard boxes or containers wasn't new (the USA used a standard container size for cargo during World War II), it wasn't until 1956 when American Malcolm McLean's converted tanker carried a cargo consisting of 58 standard shipping containers which we would recognise as such today.

This change (he named it "intermodalism") was quickly picked up by the transport industry and, in 1966, a standard size and construction of the shipping container was internationally agreed.

These are still used today; the Twenty Foot Equivalent Unit (TEU) and Forty Foot Equivalent Unit (FEU).

Over the coming decades, ships, trains, trucks and ports were converted or built specifically to handle the TEUs and FEUs.

This resulted in lower costs due to more efficient use of transport vehicles.

Far fewer men were required to load and unload and deaths and injuries were dramatically reduced.

In addition, the rate of breakages, loss and theft were all significantly improved.

In London, the employment prospects of the dockers were hit harder than other ports because the docks were not suitable for the ever-increasing size of container ships and the traffic moved to new deepwater docks in the Thames estuary resulting in a loss of over 150,000 jobs.

The lower overhead cost fuelled an exponential growth in the volume of goods being shipped around the planet.

As an example, between 1973 and 1983, the number of TEUs

moved globally tripled (four million to 12 million TEUs)[5].

## Process Improvement on the London Underground

Holborn Station was first opened in 1906 and the current configuration of escalators to take passengers from the platforms to street level was installed in the early 1930s.

Millions of people have made the journey from the trains to the surface over that time yet, in 2015, a process improvement was identified, planned and implemented using a methodology similar to those described at the start of this chapter[6].

The improvement to the efficiency of the movement of passengers from train to street level is quite simple; require passengers to stand on both sides of the escalator.

For anyone who has commuted on the "Tube" in London, this will come as quite a shock, not least because of the culturally-ingrained passive aggression deployed by those passengers in a hurry who want to walk up the left hand side of the escalators and tut tut at the out of towners and tourists who block their way.

For decades the rule has been the left hand side has been reserved for those wishing to not remain standing still while the escalator propels them upwards.

The evidence shows for the purposes of moving as many people as possible in the same direction, we've been doing it wrong.

Holborn Station is one of the deepest underground stations on

5 *http://www.worldshipping.org/pdf/container_shipping_and_the_us_economy.pdf*
6 *http://www.theguardian.com/uk-news/2016/jan/16/the-tube-at-a-standstill-why-tfl-stopped-people-walking-up-the-escalators?CMP=share_btn_tw*

the network, therefore its escalators to street level are far longer than most other stations.

In Holborn Station's case, this means fewer passengers choose to take the active option, resulting in the left hand channel having a lot of empty space whilst queues build up at train level as passengers wait patiently (this is London, the world's capital of queueing, after all) to join the right hand side and stand still on their journey to the surface.

The study and pilot exercise has conclusively proven by filling this space with standing passengers, a much greater throughput of passengers can be transported.

Specifically, the study showed 30 per cent more passengers could exit the station during rush hour by enforcing the "standing on both sides" rule (16,220 passengers compared with 12,745).

The critical length of escalator seems to be about 18.5 metres, above which far fewer people choose to walk; Holborn's escalators are some of the longest on the network, at 24 metres in length.

Those in a hurry and who were previously prepared to step up the 24 metre long escalator will have their journey slightly extended but the overall benefit will be to everyone else's journey time as the majority will no longer have to wait in line for as long.

# Chapter 3
# Automate
## The future of robotics

When we talk of robots, the first thing we must do is forget most of the science fiction in popular culture we've been exposed to.

A robot doesn't need to look like a metallic version of a human or a box-shaped lump of metal with flashing lights on legs.

Most of us are using robots in many aspects of our lives already, none of which loudly warn us of "Danger, Will Robinson".

Robotics is a function of the convergence of mechanical and electrical/electronic engineering with computer science.

As we've seen in earlier chapters, the velocity of innovation in these skills has increased significantly over the previous three decades.

In addition, an exponential reduction in the cost of the components and technology (thanks to more efficient manufacturing methods and production in lower cost economies) has made many robotic solutions justifiable to organisations building a business case.

When the cost to develop and implement a robotic solution to replace an incumbent human worker has reduced to the point where the return on investment is within an acceptable time period, that worker will be displaced.

Put simply, when the robotic solution is cheaper than, say, the combined total of three years of the employees' salaries, the job will go.

## Supersize my robot

This drive towards robotics replacing humans isn't science fiction, we are seeing this today.

If you've ever had the misfortune to find yourself in a McDonalds recently, you may have noticed a screen supplementing the cashier staff who take orders. You select the combination of menu items you require, insert payment and the food is delivered at the counter in the same way as if a human had taken your request.

This isn't particularly innovative; we've been able to order pizzas online for home delivery without having to speak to a human for some time now. But how long do we think it will be before all cashier staff are replaced by the autoteller?

The next logical step for McDonalds is to automate the cooking, wrapping and delivery of the burgers to the counter. It shouldn't be too difficult to achieve. The fries and burgers are already automatically cooked to a precise combination of temperature and time.

The addition of a mechanical device to remove the food from the stove/fryer, wrap in the paper, place in a box and deliver to the customer waiting at the counter will not be a difficult challenge to the development team at McDonalds.

Only two questions really remain; will the customers accept food cooked and delivered without human intervention and when will the development and implementation cost of the new equipment pay back against the wage bill it will replace?

There are currently around 1.5 million people working in McDonalds' restaurants. It's not hard to foresee a time in the very near future where only a fraction of those jobs will exist with the remainder gone forever.

Sure, flipping burgers in McDonalds is hardly the pinnacle of anyone's career, but it is often an entry-level job into the workforce for a sizable proportion of the adult population.

Without that first step into the world of work, the employment ladder will look very different once McDonalds' robotics programme is completed.

## Driverless Trains

Years ago, when a young boy was asked what he wanted to do when he grew up, one of the stereotypical answers would be "to drive a train".

Being the train driver or one of his assistants was seen as a prestigious, highly-skilled role for a working class man to aspire to. It was also seen as a secure job and, at certain times in history, a relatively safe job with lower fatality or injury rates than many other professions.

In fact, one of my great grandfathers was prevented from joining up for the First World War because of his job on the English railway (he was a fireman; he shovelled the coal into the boiler).

In present day India, applicants are required to complete 8 to 10 years of study before being handed the keys of a passenger train and the successful candidates are paid relatively high salaries.

The future for train drivers all over the world is bleak though.

To understand why, think back to the last time you had to change terminals at an international airport; many modern airports now have a driverless train moving passengers between the different terminals.

The technology already exists to run fully-automatic passenger trains and the additional factors which are required to be programmed for to enable this technology to scale up for more complex routes, such as a city's subway network, are unlikely to pose much of a challenge over the coming years.

Indeed, Paris has already retro-fitted driverless technology to its Metro Line 1, one of the world's oldest underground railways.

Shanghai and Delhi are already investigating the technology for their metro systems. London has had a driverless train network, the Docklands Light Railway, since 1987 with 7 lines and 45 stations now in operation.

The passengers of the London Underground must presumably think the day can't come too soon their part of the system is automated, as they suffer the almost annual series of contentious strikes by the few British trades unions with any remaining credible threat of industrial action.

Given most London Underground strikes are over pay disputes, the business case to fully automate the operation of the trains must be getting close to break-even point with every new wages settlement agreement.

Perhaps passengers are comfortable with the relatively lower speed metro lines running without a human driver, but not so with the prospect of alighting on a high speed train run entirely by computers?

Indonesia will soon learn whether this is the case as they commence work on the 150 kilometre Jakarta to Bundung line with driverless technology installed.

Safety concerns will need to be addressed before fully-utilising the technology and removing the driver from the cab. Logically though, the reaction time of a human to an obstruction on the line and the braking distance of a train whilst travelling at 200kph will never be adequate to prevent an accident caused by a collision.

## Driverless Cars

It's hard to escape the hype surrounding driverless automobiles today.

The technology currently exists to drive a vehicle safely on a road. Mining companies use it every day to move huge trucks around the open cast mines in Australia, thanks to the GPS network, where they have complete control of the roads and other road users.

Nevertheless, there are several regulatory and public acceptance issues to overcome before driverless cars will be seen on our city streets.

Major corporations are placing bets these hurdles are not insurmountable.

At the 2016 Consumer Electronics Show, Google and Ford announced they are in discussions over forming a partnership to develop autonomous vehicles. This isn't necessarily a guarantee of success (after all, you probably aren't reading this via your Google Glasses) but it will give the concept a large push toward becoming a reality.

The safety advantages promised by autonomous vehicles are almost too great to comprehend.

Approximately 1.25 million people are killed worldwide in road accidents every year, the vast majority of which would have been avoided if "human error" was removed from the list of causes.

From a 2008 report to the USA Congress[7] we can see, compared with the number of crashes caused by driver error, the other two categories (vehicle failure and environmental conditions) are negligible.

Cause of Road Accidents

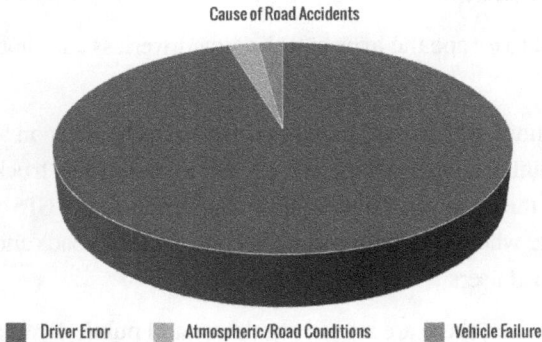

■ Driver Error   ■ Atmospheric/Road Conditions   ■ Vehicle Failure

It's not hard to imagine insurance companies would have a great incentive to avoid the costs associated with those two million accidents every year and, once the capability for autonomous vehicles has been introduced and proven, the

----

7 *http://www-nrd.nhtsa.dot.gov/Pubs/811059.PDF*

premiums to allow the owner to take back control of the vehicle would be punitive. It might become a new badge of conspicuous consumption and prestige to be able to demonstrate you can afford to drive yourself.

A Mercedes study from 2008[8] found that cars with computer assisted-braking reduced motorway accidents by 36 per cent.

The consequential reductions in loss of life, medical treatment and the human misery inflicted on surviving family and friends caused by car accidents shows the development and adoption of driverless vehicle technology is worth encouraging.

A more recent study confirms this and found a 40 per cent reduction[9] in rear-end collisions can be directly-attributable to automatic braking systems. Humans are clearly terrible drivers.

Secondary to the welcome safety dividend on offer with driverless cars, there is likely to be a significant improvement to journey times resulting from the homogenous driving style a consistent application of the road rules will bring.

Imagine a motorway where nobody sat in the middle lane driving at 10 per cent below the speed limit or a fast lane bereft of the aggressive driver flashing their lights behind anyone driving at exactly the speed limit.

As with the Holborn Station escalators, the removal of the dawdlers and the speeders will take away a lot of the causes of traffic bunching and jams.

Studies of road traffic flow have identified a phenomenon called the Congestion Shockwave where an emergency braking event

8  *http://wardsauto.com/news-analysis/mercedes-study-finds-auto-braking-slows-collision-rate*
9  *https://worldindustrialreporter.com/automatic-braking-systems-help-reduce-rear-end-crashes-by-40/?adsrc_=bluesectionarticle*

will result in a knock on effect to the following traffic which can travel upstream at a rate of around 20kph.

If you've ever had to hit the brakes in response to the cars in front doing the same but not understood what the root cause was, it was probably a congestion shockwave from miles ahead.

Driverless cars will undoubtedly result in remarkable improvements to the quality of human life with far fewer of us dying or being maimed in accidents and a reduction in our journey times.

A McKinsey study[10] from January 2016 suggested 50 per cent of all popular consumer vehicles might be fully-autonomous by 2035.

If correct, that's great news for parents of children born in 2020 or later; they can save money on expensive driving lessons!

The shockwave effect for employment is autonomous vehicles will consign millions of jobs to the history books. There are 5.7 million commercial driving licence holders in the USA, almost all of whom will be looking for alternative employment once autonomous vehicles become widely adopted.

There are even wider social ramifications to be considered too. As Uber's Travis Kalanick commented, the most expensive component of a taxi is the driver;

> "The reason Uber could be expensive is because you're not just paying for the car — you're paying for the other dude in the car. When there's no other dude in the car, the cost of taking an Uber anywhere becomes cheaper than owning a vehicle. So the magic there is, you're basically bringing the

---

10 *http://www.mckinsey.com/industries/high-tech/our-insights/disruptive-trends-that-will-transform-the-auto-industry*

cost below the cost of ownership for everybody, and then car ownership goes away."

If the cost, availability and reliability of a driverless taxi service compared favourably with the cost and convenience of car ownership, many people are likely to increasingly view car ownership as an expensive luxury.

Motor insurance companies will quickly share this view too, especially if the emerging data indicates a reduced accident rate associated with the cars driven by computers.

Most cars are unused most of the time; they sit in the garage waiting for us to make the journey to the office or a quick trip to the shops. City-dwellers get particularly low value from their cars; it's a highly inefficient asset to own for the limited utility they provide to us for most of the 24 hours in a day.

If society moves away from car ownership, far fewer vehicles will be required to make the same number of journeys and the taxis which replace our family car will have a much greater usage rate - sweating the asset, so to speak.

Consequently, the vehicle manufacturers will sell fewer units and will therefore require a smaller workforce, irrespective of the reductions resulting from the manufacturing process improvements we saw in the earlier chapter.

There is still some scepticism on when truly autonomous vehicles will be legalised in many countries but we can be fairly certain one of the first jurisdictions to do so will be a small city or island state where the disruption of the change can be planned and managed effectively.

Singapore is my personal bet, especially since the April 2016 announcement of a joint venture between the Singapore SMRT department and 2GetThere[11] to produce autonomous vehicles for use on the public transport network. Most of the major automobile manufacturers have research and development work underway with Ford and Fiat Chrysler[12] being just two who have made significant progress towards producing the first commercially-available autonomous car.

## Pilots will be joining navigators in the retirement home

Just as the highly-skilled, well-paid career of the airline navigator was made redundant with the opening up of the GPS network, the role of the pilot is going to be increasingly under threat over the next few years.

The existing autopilots installed on modern aircraft already have the capability to automatically take-off and land the plane, although this functionality is not currently mandated to be used by any commercial airline.

In fact, even when the pilot decides to "manually" take-off or land, there are automatic systems controlling aspects of the procedure.

The autopilot is flying the plane for the majority of your flight, however. The pilot is unable to control and trim the flightpath to the degree of accuracy a computer can.

To the airline's accountants, this inefficiency equals higher fuel bills and the risk of knock-on delays to schedules, which incur additional airport charges and compensation payments to inconvenienced passengers.

---

11 *http://www.smrt.com.sg/Media/Press-release/News/articleid/780/%20 News%20Releases/parentId/180/year/2016?category=Announcements%20*
12 *http://www.digitaltrends.com/cars/fiat-chrysler-seeks-autonomous-partners/*

It may be some time before we are prepared to get on a passenger aircraft that has its flightpath programmed by an engineer who then exits the plane and waves goodbye to the cabin crew (and there may not be so many of the cabin crew either!).

But there will be no such squeamishness with pilotless cargo planes. Companies such as FedEx, UPS and DHL will be among the first to apply for licences to operate autonomous aircraft to cut the cost of delivering their freight.

In the USA, cargo pilots earn between $150,000 and $200,000, so perhaps a total cost to the airline of up to $250,000 each with on-costs (pension, healthcare, annual leave, et cetera).

FedEx operate around 650 aircraft, flown by around 4,500 skilled pilots.

Once autonomous aircraft have been approved by the various aviation authorities, the business case to phase out pilots would begin to look attractive at the point that the costs of re-fitting and operating its 650 aircraft is lower than $250,000 x 4,500 x 3 years; or just over $5m per aircraft.

A lot of re-fitting can be bought for $5m.

The business case becomes particularly attractive for the 50 per cent of the fleet consisting of small, short-hop aircraft when the use of new, specialist aircraft is considered.

DHL is already using cargo drones to deliver packages to residents on the German island of Juist, in the North Sea.

There isn't even a seat on the plane for the pilot of the drone, rather he or she sits in a communication room on the mainland to control take-off and landing, presumably switching to autopilot for the journey.

Would anyone even notice if the pilot of this aircraft were replaced with a computer programme?

Who knows, perhaps, just to prove the concept internally, the autopilot has already been used for an entire journey already. How would we even know?

## Warehouses

A sector which has long been an early adopter of the benefits of IT and automation is that of warehousing and supply chain logistics.

This is an industry where managing by the metrics is a huge differentiator[13] in cost and efficiency between the best and the worst-performing players.

Inefficiencies in the warehouse and the supply chain cause knock on issues to sales, operations and stress to the balance sheet.

An inefficient warehouse is one which loses track of inventory, stores the most-frequently used items too far from the loading bay, fails to predict changes in demand due to factors such as seasonality and is unable to rapidly service requests.

Modern warehouses are unrecognisable from their predecessors of 30 years ago.

Where a team of staff in brown coats would push trolleys and ladders around, selecting items off pick lists and fulfilling individual orders using their memory and judgement of the

13 *http://www.scmr.com/article/warehousing_efficiency_and_effectiveness_ in_the_supply_chain_process/*

shortest combination of aisles and shelves, computerised picking devices now move around collecting items for multiple orders on a route calculated to be the most efficient possible.

By labelling inventory items with Radio Frequency Identification (RFiD) tags, the exact location of every item can be tracked centrally, reducing the wastage associated with misplaced stock.

These tags are the next generation of bar codes; as the name infers, they broadcast their details to distances up to 15 metres rather than requiring a scanner to be physically close and pointed directly at the tag (as with barcodes and Q codes), enabling a storeman (or robotic storeman) to quickly perform a stock take simply by moving along the aisles with an RFiD reader to log each tag as it is passed.

Logistics organisations such as Amazon (that's really what they are; a logistics organisation with a user-friendly web interface) have hundreds of autonomous small robots following colour-coded tracks on the warehouse floor, lifting stacks of shelves and moving them to their required destination to be picked from, consolidated, filled, packaged and dispatched.

An operation which would have required dozens of staff working in shifts and far from optimally is now overseen by just a few warehouse managers and maintenance staff.

## Macros

One of the current buzzwords at the outsourcing and business process optimisation conferences around the globe (and, gosh, how exciting those conferences are) is Robotic Process Automation, or "software" as you might previously have known it.

RPA has been around for years but was previously the exclusive domain of the computer programmers in the corporate data centre. Software to automate business processes was relatively expensive to design, code, test, implement and maintain as it had to be created to be generic enough to work for the entire user population of the centrally-provisioned IT systems.

Computing power is far more distributed now, and with this distribution comes a more democratic opportunity to develop solutions at an increasingly individual level to provide small packages of automation to remove the "swivel chair" aspects of some roles.

To illustrate this, imagine you are a junior insurance claims clerk and part of your job involves reading the details of a client's accident claim from the corporate database, switching to one of the free internet map and satellite image websites to take a snapshot of the location of the accident and then moving to a website with meteorology records to check the weather conditions at the time of the accident.

This information then needs to be cut and pasted into the claim record for assessment by the senior claims adjuster.

By installing a low cost piece of software which can capture keystrokes and mouse clicks, these process steps could be recorded and recreated for all future transactions.

The power of this method of automation is that it can be easily adjusted when the externally-provided websites change the position on the screen where relevant data is contained.

Previously, the centrally-provided solution created by the software developers in the corporate data centre would have been unlikely to have been prioritised above other requirements or indeed developed in a timescale fast enough to adjust to the changing external data sources.

## Journalism - an entire industry automated by the internet

Journalism has been hit hard by a perfect storm of digitisation over the last 20 years. The internet and personal devices such as smartphones have performed an exquisite pincer movement on the industry of news delivery and the profession of journalism.

With the proliferation of mobile devices, the reduction in cost of mobile data and the ability to source news from thousands of Internet sites, both domestic and international, the requirement or desire for consumers of news to pay for an unwieldy pad of paper which contains information, already out of date the moment it was printed, has decreased exponentially.

Incredibly, for an industry which is built primarily on finding new information, analysing it and broadcasting it more widely, the news industry was completely caught napping when its entire business model was destroyed within just a few short years.

Newspapers are going bankrupt or shrinking to a fraction of their scale on a regular basis as they struggle to find ways to produce a product which consumers are still willing to pay for.

To add insult to injury, the same handheld devices which killed the distribution channel for the print version of their product are also killing the method of production of their product.

Few people in modern societies don't already carry the tools of trade for a journalist and news photographer at all times during their day.

Everyone who has a smartphone, with their great quality cameras, and apps such as Twitter, Facebook, Instagram, Wordpress, and so on has the possibility to be a roaming reporter.

What chance does a professionally-trained journalist have when faced with the prospect of millions of competitors who are prepared to work for free, or at least for a couple of hundred "likes" on Facebook?

This democratising of the production and delivery of news away from a privileged few working to their, or their employers' agenda might have possibly moved freedom of speech on to a new historical high.

Opinion columnists are no longer selected by the media moguls or politically-motivated editors.

Anyone with a smartphone or computer can become a blogger and broadcast their version of the news, their opinions on the political situation or, indeed, pictures of their favourite breed of cat.

Of course, this proliferation of "information" requires the consumer to work harder to filter the inaccurate, the specious and the deliberately misleading.

Freedom doesn't come without a price tag, after all, and the real price of practically cost free access to unlimited sources of news is the increased risk that, rather than receiving pure content you are actually being sold to.

To paraphrase Jacob Silverman, there is news and then the rest is advertising.[14]

## Automation in Drug Smuggling

There's not much evidence on this as there aren't too many industry conferences where fellow drug smugglers share experiences, give awards, compete in golf tournaments and listen to keynote speakers.

However, we do have various police reports which give an insight in to the innovations used to cross some of the more lucrative international borders with expensive illicit cargoes.

Predictably, drone aircraft are already being used to smuggle drugs across the Mexican-USA border.

The USA Drug Enforcement Agency acknowledges the threat drones now pose, estimating at least 150 drone "mule" flights are made each month.[15]

The drug cartels have been infrequently using submarines or "narco-submarines" to delivery large payloads of illegal drugs between South America and the USA since the early 1990s. These have tended to be quite sizeable, with cargos worth up to $400 million.

Presumably, one of the reasons for the large size of the vessel is due to the fact it needed to be spacious enough for at least one human to pilot the submarine.

If this requirement was removed, perhaps the size (and cost) of the vessel could be significantly reduced and therefore lower

---

14 *http://thebaffler.com/salvos/rest-advertising*
15 *http://www.ibtimes.com/mexico-drug-trafficking-drone-carries-28-pounds-heroin-across-border-us-2051941*

the risk of detection.

## Sailing Drones

This is pure speculation (and subject to a pending patent and copyright application by this author....under an assumed name) but perhaps it might not be too long before we start hearing reports of illegal flotillas arriving on very distant shores.

Imagine what would be involved in designing and building a small, automatically-piloted sailing boat and wonder whether perhaps everything is place already for it to be made possible.

Take an existing entry-level sailing dinghy (the "Optimist" learners' dinghy, for example), attach a fibreglass self-righting keel to steady it in high winds, three cheap servo motors to control the rudder and trim the sails, a GPS unit, two car batteries, a solar panel and a small computer (an old Android smartphone should do the trick) with some home-written software to control the rudder and sails; a total cost perhaps between $5,000 and $10,000.

On a suitably secluded beach on the west coast of a South American country, fill the boat with a quantity of cocaine, set the coordinates for a remote beach on the east coast of Australia, tow it out to sea, raise the sails and wish it bon voyage.

Around three months later, there could be an interesting arrival on the Australian beach, although the recipient now has the not insignificant issue to overcome of how to sell the cargo without Police detection or, worse, becoming a victim of the local drug gangs who won't appreciate the new competition.

Again, the tipping point at which this becomes a viable drug smuggling solution will be subject to a finely-balanced business case of cost of materials, human resources to build the mule boat and the cost of failure (losing a shipment to bad weather, theft, detection or technical faults) and the profit margins of the successful shipments.

Perhaps it's happening already and is so successful the authorities haven't intercepted one yet?

# Chapter 4
# Offshore
## Outsourcing and Offshoring - What's the Difference?

At this point, it's useful to explain some of the terms being used.

**Outsourcing** – at its simplest definition this is paying another party to undertake work on your behalf.

As we touched upon in the introduction, outsourcing is as old as the history of villages and towns. At a domestic level, we outsource 95 per cent of everything we need in our home; food production, clothing, healthcare, education, house building, car maintenance, et cetera.

You might occasionally bake your own bread but most people outsource this task to the bakery section of their local supermarket.

Unless you're reading this whilst living on a self-sufficient homestead in a remote location, growing and spinning your own cotton, outsourcing is a huge part of your day to day life.

**Offshoring** – delivering services from an overseas location, usually from a lower cost economy; look at the country of origin listed on most appliances in your house for an example of offshored services.

**Captive** – This is an industry term used to describe an overseas delivery centre owned by the client, i.e. not outsourced but still offshored.

## Awkwardness at the barbecue

Probably with good reason, the 'offshoring' of people's jobs is almost universally viewed negatively by the general public.

Bad news stories are easy to find in the media and most people will have an anecdote they can relate about a poor experience with a call centre based overseas where a heavily-accented gentleman introduces himself as "Dave" when clearly he was actually a Deepak or Dilip.

Generally, in the west, we tend to start conversations with strangers at social gatherings in a quite predictable way; we exchange names, perhaps ask who else you know at the event or why you are attending and then comes the question which determines absolutely everything else of importance in our society, where we sit in the pecking order, which of us is superior, our background, what type of education we had, even whether we drive a prestige car or a cheap runaround....

## "So, what do you do for living?"

This is the point when things often become a little stilted for me. Several social interactions have ended quite suddenly after exchanges which follow a pattern akin to this;

Guy with a beer by the barbecue; "So what do you do for a living?"

Me; "I offshore Australian jobs to places like India and The Philippines."

Guy with a beer, probably an IT Manager; "That's terrible!

How can you justify firing all those people?"

Me; "Before I answer that, can I ask whether you consider your political views to be to the left or right?"

Guy with a beer; "Slightly to the left."

Me; "So, you believe it desirable for some wealth distribution from the one per cent who have the most money?"

Guy with a beer, suspiciously; "Yes, of course."

Me; "So why would you object to lifting people out of poverty in a less well-developed country?"

Guy with a beer, with a scowl on his face; "Erm, I don't. But what about the Australian workers?"

Me; "Well, if Socialism only applies to your own country, we could call that National Socialism, couldn't we? Now where have I heard that term before?"

At which point, my long-suffering wife usually drags me away to talk to someone else about sport.

As facetious and risking a punch in the face as the above fictional exchange might be, it is valid to investigate the ideological arguments which can be made around offshoring.

Hopefully, much of this book explains offshoring is only the final of three inevitable steps and there's little point in trying to resist this cycle of elimination of unnecessary processes, automation of tasks (where economic to do so) and offshoring of work to lower cost economies which has been gathering pace for three decades now.

The toothpaste is unlikely to ever be pushed back into the tube, not without a massive reversal of the terms of trade between sovereign states and a cessation of most research and development.

The "problem" we are faced with in the west is that we are all so incredibly richer than everyone else in the world.

In western countries with generous government benefits, social medical care and welfare systems, even those on the very lowest rung of the socio-economic ladder are edging into the global richest 1 per cent, which certainly explains the current migrant crisis which is impacting Europe —as long as this massive imbalance exists, to paraphrase Milton Friedman; we can have a welfare state or we can have open borders but the two concepts are mutually-exclusive.

It might not feel like it but, if your annual income is US $36,000 or more, you are in the wealthiest one per cent[16] of the world.

YOU are the one per cent the Occupy Wall Street movement were complaining about!

This is you if you earn US$36,000 or more

---

16 *http://data.nber.org/oww/*

## The best welfare programme is a job

If, like my offended fellow barbecue guest, you are on the left of the political spectrum, perhaps I can persuade you that providing goods and services from Third World countries such as India or The Philippines is sharing wealth far more effectively than any aid programme or charity could ever hope to achieve.

In the last three decades, the middle class in India has grown[17] from seven per cent to 46 per cent, lifting 431 million people out of a life of extreme deprivation.

That's a middle class 25 per cent larger than the entire population of the USA and 25 times larger than the entire population of Australia.

This is almost completely due to the new economic opportunities provided by delivering goods and services to Western countries which were previously provided from within those Western countries' borders. In the words of Ronald Reagan, "... the best social (welfare) program is a job".

The corollary to the wealth distribution argument is if your personal ideological view is that capitalism has done more good than any other system humans have previously experimented with, you can view offshoring as the free market moving supply to those best-suited to deliver the services.

It will be entirely logical to you that if a team of accountancy graduates in Pune can process your company's accounts payable function just as well as the previous team in Houston could but at a discount of 30 per cent in salary costs, they

---

17 *http://ecell.in/eureka13/resources/tracking%20the%20growth%20of%20 indian%20middle%20class.pdf*

should be given the contract to do so.

Obviously, appropriate checks should be made to ensure the delivery of those services is ethical, sustainable and doesn't add any risk to the company, but those checks should be made of any supplier before engaging them regardless of delivery location.

Unions are obviously particularly vocal in their objection to any suggestion their members' jobs might be made redundant and performed from an overseas location.

For an example of the special pleading unions attempt to prevent the unpreventable, there is a document in the public domain[18] which is a submission from 2009 by the Australian Services Union to the Australian Federal Senate Committee Inquiry on the Environment, Communication and the Arts.

In the 10 page document, there are calls to force organisations to disclose which services were provided offshore, mandate government procurement departments to only use domestic sub-contractors (regardless of whether these are good value for the taxpayer), provide tax breaks for domestic businesses and the submission also raises unsubstantiated concerns about the ethics and standards of the overseas employers.

Interestingly, the one question the submission fails to address is why buying services delivered from overseas is such a terrible concept, the assumption being that any reader will automatically share the common view offshoring must be prevented.

Preventing offshoring from occurring is actually extremely

---

18 *http://www.asu.asn.au/documents/doc_download/257-2010-keeping-jobs-from-going-offshore-protection-of-personal-information-bill-2009*

feasible by use of trade restrictions such as import tariffs.

These are set and constantly tweaked based on market conditions and international trade agreements.

In most Western countries, changes are made to international trade agreements at Prime Minister or Presidential Cabinet level, not requiring parliamentary ratification.

The Prime Minister simply needs to have the backing of his or her Cabinet Ministers to slap a punitive tariff on services provided from overseas.

The investigation and application of these tariffs would be similarly straightforward to undertake as the payments would appear in company's audited accounts each year end.

If the process to prevent offshoring is so simple, one must question why it hasn't happened so far? We will discuss this further in the next chapter.

## The history of offshoring

For the sake of consistency and brevity, we will continue to use India as the area of study as it was the first country to deliver sophisticated services from offshore and the location most people associate with the term.

The detail would have been very similar if we had examined China, The Philippines, Mexico, Poland or any other number of low cost locations now providing these services.

Just as organisations emerged during the late 1970s and 1980s in countries such as the USA, UK and Australia to deliver "back office" services, mainly in the emerging IT sector but also in areas with a high headcount and manual activities such as payroll processing, similar companies were delivering the same services to domestic customers in India.

In many cases, these were spun-off internal departments from existing large corporations.

Wipro, for example, now one of the world's largest IT and business services companies, was first registered in 1945 as "Western India Palm Refined Oil Ltd".

Today, the IT departments of companies such as General Motors are run by an Indian palm oil refiner.

In the mid-1980s, Ross Perot, CEO of USA-based EDS and N. R. Narayana Murthy, CEO of India-based Infosys probably didn't realise they were about to become major competitors of each other within the decade.

A series of technological and geo-political changes would bring this about.

Throughout the 1980s, the Indian suppliers were beginning to win contracts for software development for overseas clients.

Software development at that time was a very different beast to the present day; the analysts and programmers would write their allocated packages of code on monochrome, text-only "dumb terminals" directly attached to a physically-huge (as opposed to having huge processing power) mainframe.

Code written during the Indian day could be delivered for customer testing in time for the start of day in New York over a dial-up telecommunications link at 9.6kb (over two thousand times slower than most of today's domestic internet connections).

The lower cost programmers in India may have been an irritation to Ross Perot and EDS but they would be unlikely to be causing him sleepless nights...yet.

Things started to change fast once the prices of international telecommunications started to fall.

Not only did the cost of the international telephone call begin to fall dramatically but so did the availability and cost of T1 leased lines (dedicated data telephony links between two buildings – in effect, a constantly open telephone line).

The communication limitation requiring programmers to work "offline" was removed and far more interactive communications could be had with the IT systems they were developing and testing and the customers whose requirements were being coded.

The labour arbitrage between the Indian and western workers resulted in up to 40 per cent savings in wage costs, depending on scale and availability of skill set.

As the telecommunications restrictions were removed and the Indian services companies became more sophisticated in their approach to winning and retaining business in the West, the scale and scope of the services expanded to more than just tightly-defined programming work to include many other business processes.

## Reputational Issues

Everyone has a bad experience anecdote resulting from a poor customer service conversation with a worker in an overseas call centre.

In fact, it's usually the first objection offered to me when I suggest that we might offshore a department or even just part of a process.

These anecdotes are generally the result of bad decisions by the management of a company who had been blinded by the potential cost savings and didn't focus strongly enough on the risks of relocating a customer-facing operation to another country, linguistic accent and culture.

In addition, the temptation of marketing departments to push hard sales techniques to consumers using cheap labour and high-pressure cold calling has had the effect of training many people to reject unsolicited calls from numbers starting with "+91" (India).

As objectionable as these sales pitches are, one assumes they are analogous to the Nigerian internet scam emails; even though the success rate of these individual communications is miniscule, the payback for those which are successful funds the entire operation.

When I advise clients on what aspects of their workload are suitable for offshoring, client-facing communications is never on the list.

Conversely, an internal IT department's help desk, however, is very suitable.

For most organisations, the quality of service for the internal IT department the balance sheet can support isn't as high as that which the same organisation wishes to offer its clients.

If you don't like speaking to an Indian to have your corporate desktop computer fixed, you might want to consider applying for a job with Goldman Sachs instead.

## Captive or free?

When a Western company determines there is value in having some services delivered from offshore, there are broadly two ways they achieve this.

The most common approach is to issue a tender to a group of suitable suppliers based on market research of their capabilities, reputation and size relative to the client organisation.

The scope, scale and price of the services are described in a commercial contract and the ongoing relationship is managed formally using performance metrics and regular governance meetings at all levels of both organisations.

The staff working for the client might be dedicated to the account or shared across other clients or a blend of the two.

The second, less frequent, strategy is to set up a dedicated operation overseas, staffed with a newly-recruited team and managed (usually) by expatriate managers.

The industry term for this arrangement is a "captive" centre.

Many advisors, including me, tend to view "captives" with slight disdain due to several key disadvantages over the more traditional model.

Firstly, Indian university graduates are as discerning and selective as their Western counterparts; those with a first class degree from a top university are going to expect to work for the most prestigious employer brand with the greatest depth and breadth of career opportunity.

There then exists a very sophisticated pecking order which

matches the quality of candidate to the perceived quality of employer.

Indian companies such as TCS or Infosys with a workforce of several hundred thousand employees and hundreds of clients across dozens of industry sectors can offer so much more than a "captive" centre for an internationally-unknown company such as the Australian and New Zealand bank, ANZ (not to pick on ANZ, particularly).

Therefore, the only tangible way a company such as ANZ can attract better talent is to offer higher than market salaries, benefits and working conditions, somewhat defeating one of the main objects of the exercise of offshoring.

Secondly, there's a slight arrogance and immaturity to the conclusion that the best way services can be delivered from overseas, despite the existence of many companies in that country already employing thousands successfully delivering those same services, is to start from scratch and do it oneself.

Think of the effort and resource involved in registering a legal entity, sourcing and leasing a building, recruiting a team and managing the delivery compared to simply paying a service fee to cover the incremental increase to the existing capacity of a local company.

There's also the question of who will fill the local management roles.

Once again, there's a misplaced superiority complex bordering on the final years of the British Raj to think that none of India's one billion population can manage a team of accounting clerks or IT support staff and, instead, a Western manager must be flown in on an inflated salary for a fixed term to be their man or woman on the ground.

Consider also, the low probability this modern day Clive of India is actually even the best person for the job or simply the only one who put their hand up for the "hardship posting".

Without wishing to be too uncharitable, it's highly unlikely the person who agrees to move from London to Mumbai is ever going to be the best potential candidate.

Examining the way local services providers recruit in countries such as India illustrates the uphill battle a "captive" centre has to fight to secure talented new staff.

All of the main offshore services organisations invest heavily in teams who regularly visit the local university campuses looking for the best students and encouraging them to apply for their graduate intake programmes.

Often, job offers have been made some time prior to the final examinations, securing the next wave of graduates.

The city of Pune in the state of Maharashtra, India is sometimes referred to as "the Oxford of the East" due to the number and quality of local universities it hosts.

These institutions are partially-sponsored by and supply graduates to several large companies based at the neighbouring Hindewadi business parks.

There are approximately 30 companies with campuses there including some of the largest offshore services companies such as Infosys, Wipro, Tata, Tech Mahindra, Satyam and Cognizant, employing almost half a million staff in just one location.

This is clearly not the location to set up a captive offshore centre and have any hope to compete for quality local talent.

A 2011 study[19] found 60 per cent of captives fail to live up to expectations.

It's unsurprising then, that the offshoring trend is moving away from captives and to a consolidation of the large professional services organisations.

## The Risks Associated with Offshoring

One of the first steps I undertake when commencing projects to offshore functions for clients is to facilitate a risk workshop to identify what new risks and opportunities the proposed project brings to the organisation and explore the possible actions available to mitigate these.

Most of the risks are common across all offshoring projects so it is perhaps useful to describe some of these below.

### Reputation

Will public knowledge of the project damage the company's brand with its customers or the community in which it operates?

In some cases, the answer to this question is "undoubtedly, yes."

If the organisation is one of the largest employers in a town or region or where a brand based on community engagement has been carefully nurtured, perhaps offshoring is not the best choice.

---

19 *http://www.cio.com.au/article/373187/offshoring_captive_center_rises_again/*

Of course, if the competition has no such scruples and can lower their prices as a consequence of the cost savings, it might be time to readdress the value of that brand.

In the vast majority of cases though, the customers welcome the potential for cost savings to be passed on in the form of reduced prices. If this last statement doesn't seem particularly credible to you, take a look at the label in the clothes you are wearing and see if there is a single item with the words "made in the USA" (or UK, Australia, et cetera).

## Service Quality

There is always inherent risk when transferring the location from which a service is delivered and switching to a new group of staff.

The new team are unlikely to be as experienced as the old team and certainly not in the nuances of the specific process that is being undertaken.

There are several mitigations to this, the first of which is to ensure the project to transition the services focuses hard on measures to check the quality of the service is being tracked and maintained.

This is the "how" rather than the "what" of the process.

The real key to successful offshore delivery is governance.

This is where the client invests as much (or more) management time and resources to constantly checking the correct metrics are being tracked and plans are being actioned to continuously improve the delivery quality of the offshore team as they did previously before offshoring.

Whenever I hear of an offshore service with a reputation for poor quality, my suspicions immediately fall to the management of the service by the client rather that capability of the offshore services company.

I've met with executives of major corporations who have complained to me about the performance of their offshore supplier but don't see the incongruity of their answer "never" when I ask them when was the last time they visited the delivery centre and met with their peer within the supplier's organisation.

We often get the outcomes we deserve....

## Protection of Sensitive Data

Quite rightly, the security of sensitive data such as personal information or intellectual property is paramount to successful organisations.

Providing access to this information to an offshore team for processing purposes does increase the risk of loss of control of who can see data.

The offshore industry addresses these concerns by strictly-controlling access to the dedicated rooms from which a client's services are provided, removing the USB functionality of the desktop computers the staff use and enforcing a policy of no camera phones, USB drives and printers in the offices.

For banking and finance customers, several of the suppliers use specialised "data masking" technology which prevents an individual from seeing all of the digits in a credit card number, for example.

In Australia, the financial regulator, the Australian Prudential Regulating Authority (APRA) examines and audits all proposed offshore projects and the mitigations described above are regularly given approval, subject to the required evidence of compliance.

A residual risk remains that a malicious employee in the Indian office could memorise some limited details for use later.

However, this is a risk organisations must face already though with their existing employees and the mitigations are similar regardless of location.

## Wage Inflation

One of the risks sometimes offered to me as a counter to the suggestion a department would make a good candidate for savings, improved quality and control is that inflation is "racing away" in countries such as India, China and The Philippines and the cost benefit will soon not be worth the disruption of switching delivery locations.

None of us can be sure what the inflation rate of a particular country might be over an extended period but the chart on the next page plots out how the Indian and Australian salaries might differ over time if the current inflation rates in those two countries continue.

In this example, an Australian accounting student graduating in the late 2060s might have a fair chance of entering a career with a future.

Of course, this assumes that processing of supplier invoices and allocating costs to the correct general ledger codes hasn't been fully automated by then (which, of course, it will have been – many companies have already completed this).

Clearly, the wage inflation disparity between the low and high cost economies isn't going to prevent jobs from being offshored.

**Wage Inflation Australia v India**

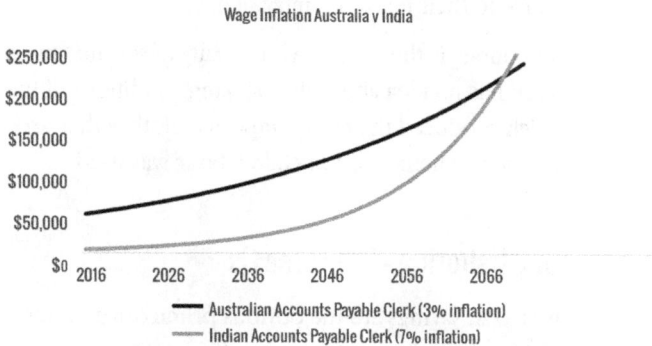

- Australian Accounts Payable Clerk (3% inflation)
- Indian Accounts Payable Clerk (7% inflation)

*(Assumptions; Australian salary in 2016 - $48,000, Indian salary in 2016 – 200,000 Rupees, exchange rate – 48 Rupees to the Dollar)*

## Sweat Shops and Exploitation of Workers

Perhaps the most desperate and ill-informed claim in the risk workshop is that the new team in India will work in "sweat shop" conditions or will be exploited in some other way.

As we've explained earlier, the middle class population in countries such as India has grown exponentially in the last couple of decades.

The wage inflation rate of seven per cent should be a taken as a solid measure the workforce isn't entirely on the back foot in negotiations with the "greedy CEOs".

Another good indication is the rate of car ownership in India.

This has multiplied by a factor of 18 since the 1960s[20] compared to a factor of four in countries such as the USA.

A quick glance at the car park in the outsource service providers' campuses across India suggests the workers who are running your employer's IT department have a good disposable income relative to their fellow countrymen.

The irony of course, is the person who usually raises the "sweat shop" concern has no idea about the working conditions of the factories which produced the cheap imported clothes they are wearing or, worse, whether or not child labour was used.

## Drivers for Offshoring - The Three Cs

In most cases, cost savings are the obvious prime driver in the move to transfer the delivery of services offshore.

There are plenty of embarrassing examples in the press where savings haven't been achieved and the quality of services have been dire but, as a ratio to the volume of work offshored, these are thankfully rare.

The reasons why offshoring can fail are numerous and the worthy subject of a separate book with the working title of "Clients of offshore services usually receive the outcomes they deserve".

Cost savings are considered the "table stakes" for offshore arrangements. Over a three year period (most deals are between three and five years in duration) the cost of the project to change delivery locations added to the ongoing cost of delivery should be cheaper than the previous "onshore"

20 *http://www.econ.nyu.edu/dept/courses/gately/DGS_Vehicle%20 Ownership_2007.pdf*

equivalent costs.

In most cases the savings will be at least 20 per cent.

There are two other key benefits which are usually considered, however; Capacity and Capability.

Capacity is often a major consideration for organisations struggling with the challenges of flexing the workforce up and down to meet with demands in their business.

For example, if a bank is planning a major IT implementation, there will be a requirement for many IT staff to work on a project which might take about a year to implement.

The choices available to resource this project are to use "permanent" staff who would then need to be redeployed or made redundant at the end of the project, contract staff on expensive day rates or an offshore sub-contractor's staff working in a lower cost economy.

In addition to the labour arbitrage benefit, the bank has a much larger workforce available to choose from in a country such as India or China.

This also goes some way to solving a problem which can often occur that two major corporations might commence similar IT projects in the same location and find themselves competing for the same talent in a small market.

This happens more often than one might think.

When one bank implements a new innovation (contactless debit cards, for example), their competitors all have to rush to follow suit.

The third of the three Cs is Capability.

In its simplest form, this is where the client organisation can access skills and knowledge which it doesn't have in-house such as particular specialist technical knowledge.

However, there's a wider capability benefit to outsourcing functions to large specialist services providers which comes from the scale of the services these companies deliver to their clients; a corporation with $1bn of revenue might have 20 accounts payable clerks while an offshore Business Process Outsource services company might employ 2,000 people performing the same role.

This economy of scale can enable new automation, tools and techniques to be developed which the client organisation would never have considered cost-effective to develop for themselves but can now benefit from.

## The dirty secret of offshoring; it can be better quality

Mainly because of the high-profile offshore project failures, the public perception is service quality takes a rapid dive when functions are moved to offshore delivery.

This very definitely can and often is the case, especially where the client's management have sleep-walked into a poorly-written contract or failed to apply an appropriate level of diligence to monitoring and managing the outcomes they've signed up to.

Things don't have to be this way and there are many offshore teams which have been quietly operating at quality standards significantly higher than the previous onshore team performed.

In my view, a key factor in this phenomenon is the clarity

required when services are offshored under a commercial agreement between two companies.

It's a sad fact the first time a department finally gets around to that annoying task of documenting its business processes and setting performance targets to measure quality is when the Outsource Transition Team arrive on an Air India flight.

It's all too late then as the next step after documenting the "what" and "how" is to transfer the knowledge to perform the work so that they can take it back and train the "who" - their colleagues in India.

The old adage is true; you can't manage what you don't measure.

If you're a department head who wants to protect the team from the threat of offshoring, you're unlikely to do so by suddenly becoming the world's greatest advocate of management by metrics but not knowing how your performance compares with industry benchmarks is a sure way of quickly losing the argument when the offshoring feasibility study is proposed.

## An Indian Offshoring Case Study

One of my recent clients, a mid-sized Australian company, has embraced the offshoring and outsourcing model to an extent which few Australian organisations have so far.

This was initially out of financial necessity; the financial crisis of 2008 hit their clients hard and, as profit margins became squeezed, cost savings were sought in ways previously deemed to be unpalatable.

Subsequently though, after the benefits of offshoring had been proven, the drive to find more opportunities to send departments and processes to India became mandated from the top of the organisation down.

As is often the case, the first team to be offshored was the IT department.

The support and maintenance of the multiple business software applications, networks and hardware was transferred under a five year contract to a specialist IT services provider in India.

Where a member of staff would have previously called a local IT technician to come to their desk to resolve an issue, they would now speak to one of a team of staff in India.

The majority of the teams of support staff were also based out of India to a ratio of roughly 80 per cent.

Culturally, this came with some challenges, not least a tangible undercurrent of racism from the client's staff.

In fact, a few months into the contract, two of the client's staff were formally disciplined for the inappropriate way in which they spoke to the Indian help desk staff.

The Human Resources meetings where we replayed the recordings of a member of staff racially abusing a complete stranger were particularly unedifying for all participants.

A word of advice when you're speaking to someone more junior than you on the telephone, regardless of their location or ethnicity; imagine having to listen to the conversation replayed later in front of an audience and then modify your language and tone accordingly.

The IT offshoring project resulted in operating cost savings of

30 per cent from the previous year, not including the expensive office space in Sydney which was subsequently freed up for other use.

The project resulted in around 85 jobs leaving Australia, unlikely to ever return.

Buoyed by the success of their first offshore project and, with the confidence of experience, the company then asked me to help with offshoring some of the "back-office" financial and human resources functions.

An offshore services provider which specialised in providing payroll processing, accounts payable/receivable (the processes to pay suppliers and invoice clients) and administering general human resources paperwork such as contracts of employment was selected.

Despite this being their second foray into offshoring, there was a high degree of nervousness around the risks of moving the payroll processing function; after all, failing to pay your staff correctly or on time is one of the quickest ways to go out of business.

The "dirty secret" of offshoring was borne out to be true again though as we learned during the transition project that many of the payroll processes we were documenting had never previously been formalised and the key metrics to measure accuracy and timeliness of the payroll function were flawed or didn't compare well with the industry benchmarks.

Within three months of transitioning the payroll function, the Indian supplier was over-achieving the previous targets and, within six months, was meeting the industry best practice for payroll accuracy.

Previously, the completed file with all of the pay details for the employees would miss the bank's cut-off deadline on a regular basis and extensions had to be sought.

Within a few months of the move to India, the file was being delivered with hours to spare.

It would be misleading to suggest the offshore supplier was completely responsible for these improvements, but by taking the overhead of running the day to day activities away from the client's payroll manager, she was able to direct her focus to solving the underlying issues with the processes such as managers and staff not respecting cut-off deadlines to requests to change terms and conditions or submit approved timesheets.

The cost savings for this second wave of offshoring were also approximately 30 per cent from the previous year and, in addition to the labour cost savings, an entire building was able to be released for sub-letting once the 120 jobs had moved to India.

The next move the client made was to re-organise their business by looking at the many "back office" functions that were being performed at their many local branches.

Typically, each branch would have at least two people who undertake a range of activities which fell in to the category of "Finance and Administration".

These ranged from responding to 1970s-type management requests like picking up the dry-cleaning for the most senior person on the site to raising purchase orders and paying invoices through to accountancy work such as the reconciliations for the month-end processing.

Remembering the order of preference, these roles were

eliminated (the boss can collect his own damn dry-cleaning), automated (we have standard reports that can be viewed for month-end) or offshored.

The client now has a rolling programme of scrutinising the work performed throughout the organisation to question whether it really actually needs to be performed, whether there's a technology solution if it is still required and, if not, can the function be packed up to be performed in a lower cost economy?

## What else can be Offshored?

The case study above shows there's a wide range of services which can be provided from the low to high cost economies for just one organisation.

The list of functions currently being successfully sourced from overseas is far greater. The following are just a few of the mature, tried and tested offshore services which many companies are using.

### Recruitment

Recruitment activities which are regularly delivered from overseas include tasks such as sifting and sorting of candidate applications and profiles, arranging interviews, taking references and completing all the arrangements (such as allocating a desk, computer, mobile phone and setting up payroll) required for the successful candidate to commence work.

## Legal

Shakespeare's character Dick the Butcher famously suggested ".....let's kill all the lawyers".

Whilst many of us can sympathise with Dick's sentiments, in the absence of this being allowed, we can offshore the lawyers instead.

Member of the European Parliament Dan Hannan terms countries such as the USA, Canada, UK, Australia, New Zealand, India, et cetera as "The Anglosphere".

These countries all share a legal system based on the 800 year old English Common Law.

There are obviously local case law and legislative variations but the similarities are far more material than the differences in their impact to a legal graduate trying to learn the legal code of a new jurisdiction.

Contract law, for example, is not particularly dissimilar across many of these Anglosphere countries and therefore those repeatable tasks such as reviewing clauses and schedules which would have previously been handed to the junior lawyer in a city law firm in London are now being undertaken in India.

Quasi-legal tasks such as the ongoing monitoring of contract compliance (managing key milestone dates and deliverables or ensuring pricing discounts based on volumes are applied) are also obvious candidates to be undertaken from lower cost locations.

## Executive Assistants

Above a certain grade and pay level in most organisations, the senior managers benefit from the assistance of an executive assistant to organise their appointments, travel plans, field their phone calls and many other activities we mere mortals somehow manage to undertake independently.

Depending on the nature of these services, there's very little reason why this function can't be performed remotely and this is a small but growing service offering from countries with a good reputation for customer service such as the Philippines.

In some cases, for the owner of a small business for example, the offshore executive assistant is performing a job which might never have existed previously because the cost of employment in the West would have been prohibitively expensive.

## Travel Services

There is no compelling reason why the researching, booking and making changes to schedules for corporate travel needs to remain in the high cost economies.

All of the major airlines and travel companies now have online booking systems which can be accessed remotely.

If your employer uses one of the main corporate travel booking services, it's highly likely there is already a team offshore supporting and maintaining the bookings you make in their system.

## Corporate Design

Somewhere in the vicinity of the marketing department is often to be found a group of people which will include graphic design artists, web designers, Photoshop experts and other people with other, related skills.

Most of their working day will be spent looking at a computer screen (always a very big screen and usually made by Apple) and moving a mouse a lot.

There is little which prevents these roles being performed from a lower cost economy.

## Animation

One sector which has experienced more than its fair share of disruptive change over recent years is the animation industry; the production of cartoons for TV or cinema release has been revolutionised by both advances in digital technology and the availability of cheaper labour overseas.

Disney's 1937 classic Snow White and the Seven Dwarves employed 402 artists[21] based in California during its production.

Glancing at the credits of a modern animated movie such as Kung Fu Panda 3 shows that Chinese–based studios such as Oriental Dreamworks[22] a subsidiary of the Disney competitor, Dreamworks Animation, have taken on a large slice of the present day equivalent work.

Animation is still a very labour-intensive process but the labour now no longer needs to be based in a high cost economy.

---

21 *http://backlots.net/2012/11/28/disney-production-process-and-innovations-in-animation-technique-in-snow-white-and-the-seven-dwarfs-1937/*
22 *http://mediadecoder.blogs.nytimes.com/2012/02/17/dreamworks-animation-forms-studio-with-chinese-partners/*

## Everyone Else Who Uses A Computer And Doesn't Talk Much

This is the catch-all category for all those other jobs I could list.

When I walk into a large open plan office, I look and listen to the activity; if the overwhelming noise is of keyboard strokes rather than human voices, it's a good clue that much of the function being performed isn't location-dependent.

If your job requires a lot of "screen time" and little human interaction, it's probably only a matter of time before an offshore services company will develop a "centre of excellence" to perform it.

# How big is the threat of offshoring?

Unsurprisingly, there is no central register of jobs which have been removed and are now provided from overseas. Various estimates are available however, mainly collated by looking at news articles when the initial deal is signed between the two companies.

This is obviously flawed in several ways. Firstly, not every offshore contract is reported upon.

Secondly, there's no guarantee the reporting has accurately captured the scale of the deal and may have over or underestimated the impact.

Lastly, successful offshoring contracts have a tendency to be extended and have an increase in scope, neither of which is likely to trigger the radar of the few remaining employed journalists. Nevertheless, some estimates of scale which are available include the following;

## Australia

An Australian federal government report identified 51 categories of work which it considered at risk of being impacted by offshoring.

The number of workers in those professions at the time totalled over 10 million people, or nearly half the entire population of Australia.

The same report repeated an estimate by the Financial Services Union (FSU) that 15,000 Australian jobs were offshored in 2007.

## Occupations potentially affected by offshoring[23]

| | |
|---|---|
| Engineering managers | Legal Professionals |
| Information Technology Managers | Economists |
| Sales and Marketing Managers | Chemists |
| Urban and Regional Planners | Geologists and Geophysicists |
| Policy and Planning Managers | Life Scientists |
| Journalists and Related Professionals | Financial Dealers and Brokers |
| Authors and Related Professionals | Medical Scientists |
| Branch Accountants and Managers (Financial Institutions) | Customer Service Managers |
| Environmental and Agricultural Science Professionals | Architects and Landscape Architects |
| Financial Investment Advisers | Quantity Surveyors |
| Computing Support Technicians | Bookkeepers |

23 *http://www.aph.gov.au/About_Parliament/Parliamentary_Departments/ Parliamentary_Library/pubs/rp/RP0708/08rp03*

| | |
|---|---|
| Other Natural and Physical Science Professionals | Cartographers and Surveyors |
| Other Managing Supervisors (Sales and Service) | Credit and Loans Officers |
| Secretaries and Personal Assistants | Civil Engineers |
| Mechanical, Production and Plant Engineers | Electrical and Electronics Engineers |
| Advanced Legal and Related Clerks | Insurance Agents |
| Mining and Materials Engineers | Accountants |
| Desktop Publishing Operators | Keyboard Operators |
| Accounting Clerks | Auditors |
| Marketing and Advertising Professionals | Payroll Clerks |
| Computing Professionals | Bank Workers |
| Mathematicians, Statisticians and Actuaries | Librarians |
| Money Market and Statistical Clerks | Insurance Clerks |
| Business and Organisation Analysts | Switchboard Operators |
| Other Business and Information Professionals | Telemarketers |
| Medical Imaging Professionals | |

## USA

Conflicting studies have suggested between 150,000 and 300,000 American jobs[24] are lost each year to offshore competition and U.S. Department of Commerce data reported that "U.S. multinational corporations, the big brand-name

---

[24] *http://www.economist.com/news/special-report/21569568-offshored-jobs-return-rich-countries-must-prove-they-have-what-it-takes-shape*

87

companies that employ a fifth of all American workers… cut their work forces in the U.S. by 2.9 million during the 2000s while increasing employment overseas by 2.4 million."

## UK

A 2011 study by the World Trade Organisation[25] suggested 6,750 jobs were lost in the UK in one year (2005).

More recent data is difficult to source for the reasons stated earlier in this chapter but it probably a fair assumption that the number of job losses attributable to offshoring would sit midway between that of Australia (15,000) and the USA (150,000).

---

25 *https://www.wto.org/english/res_e/booksp_e/glob_soc_sus_e_chap1_e.pdf*

# Chapter 5
# Don't look to the Government for help

Over the coming decades, if unchecked, we will witness vast swathes of the job market in the West disappearing through elimination, automation and offshoring in a continuation and acceleration of a trend which is already two decades old.

School leavers in high cost economies such as the USA, UK, Australia in 2030 will be faced with a dramatically-reduced landscape of employment opportunities and those surviving professions will have a much higher standard requirement to join as the entry level jobs will either not exist or be performed elsewhere.

A society with a high population of unemployable and disaffected youth is not a pleasant prospect to contemplate.

Faced with this clear and present danger to the economy and law and order, what is likely to be currently exercising the highly-paid brains of the policy-makers in our respective governments?

Two questions might leap to mind;

1. Is there anything governments can do to stop the destruction and exodus of jobs?

2. Even if governments could stop it, should they and what are the alternative courses of action?

As we saw in Chapter 4 - Offshore, preventing services from being provided from overseas could easily be achieved by implementing trade restrictions.

Regulations, quotas, import tariffs and other government mechanisms to throttle imports from entering their market freely have always existed and are constantly being tweaked to respond to threats and opportunities, as the "Roquefort War"[26] illustrates.

In its final days, the Bush administration imposed a 300 per cent duty on Roquefort, in effect closing off the U.S. market.

"Americans, it declared, will no longer get to taste the creamy concoction that, in its authentic, most glorious form, comes with an odour of wet sheep and veins of blue mould that go perfectly with rye bread and coarse red wine.

The measure, announced Jan. 13 by U.S. Trade Representative Susan C. Schwab as she headed out the door, was designed as retaliation for a European Union ban on imports of U.S. beef containing hormones. Tit for tat, and all perfectly legal under World Trade Organization rules, U.S. officials explained." (*Washington Post 29th January 2009*)

If the imposition of punitive trade restrictions against offshore service providers is straightforward and legal under the WTO rules, why aren't our governments protecting our jobs, especially considering our politicians fear losing their jobs at election time if they displease us?

The answer becomes clear when we look at the consequences of unilaterally limiting international trade.

---

26 *http://www.washingtonpost.com/wp-dyn/content/article/2009/01/28/AR2009012804071.html*

No sensible government wants to trigger an international trade war.

They are fully aware the defining trend of the last three decades is of the loosening of international trade restrictions and this has brought many material benefits to their voting demographic.

It would be a brave administration which tempts fate by risking returning to the bad old days of closed markets.

Their voters have become accustomed to driving prestige German cars, browsing Facebook on Korean-manufactured smartphones, eating imported summer fruit regardless of the season and wearing clothes that are so cheap as to be practically disposable after being worn just a few times.

If a government unilaterally imposes a trade barrier of any real meaning (rather than just a spiteful tax on exquisitely-tasting sheep's cheese) the rest of the world will reciprocate.

Consider the impact on a modern Western country if it simultaneously had to cope with the closure of foreign markets for its exports of goods and services AND had to switch a large proportion of the workforce back to performing manual and menial roles which were deemed to be undesirable 20 years ago.

In fact, regardless of the impact on the economy, imagine the effect such a backward step would have on the national psyche as that country had to resume manufacturing their own clothing, furniture, plastic toys, et cetera.

Regardless of the very real threat to the longevity of the life of entitlement, status and luxury experienced by our legislators, there is also the threat of the law of unintended consequences;

the bear trap which most government legislation seems predestined to fall head first in to.

Governments of all political hues fear few economic phenomena more than rampant inflation and central banks are under intense pressure to not create situations which might unwittingly revert to a highly-inflationary environment.

Countries which have experienced not just high inflation rates but hyper-inflation, such as Germany, have a national memory and phobia of the condition.

The stories of the effects of the three years (1921 to 1924) of hyper-inflation during the Weimar Republic era still haunt the policy makers in the German Bundesbank, and were the cause of much angst and pause for thought when they were agonising over the correct action to deal with the currency crisis in the aftermath of the 2008 financial meltdown.

One of the main reasons we've not experienced high inflation in recent years is the relative freedom of international trade.

As we explored in Chapter One, in crude terms, Western economies have been importing deflation by buying goods from lower cost economies. Prices have been falling for most consumer products for years, having the effect of keeping downward pressure on local wages. It's hard to ask for a pay rise when the CPI (Consumer Price Index) is showing barely two per cent inflation and your employer's margins are under pressure from cheap foreign imports.

In addition to cheap imported goods, "importing" workers from low cost economies also has a deflationary effect. There's a reason why immigration and tax laws are the most frequently amended of all national legislation; governments realise that these are two of the more effective levers they can pull to manipulate the economy and seek public approval.

This is not without problems, though, as successive British governments are discovering as recent loosening of immigration restrictions has impacted local services such as healthcare and welfare and kept wage inflation low, both of which have contributed to negative public opinion culminating in the vote to leave the European Union (EU) on June 23rd 2016.

Ultimately, immigration controls are the main reason why jobs are offshored to India; if companies were able to fly in hundreds of thousands of Indians to a high cost economy and pay them at Indian rates, of course they would do so.

The fact they are unable to do this has driven the push to find technological solutions to move the roles offshore.

Industries with a large proportion of workers who are union members are, counter-intuitively, most at threat.

Unions exist for two main purposes; to protect the health and safety of their members from exploitative employers and to fight for improved pay and conditions.

Without wishing to downplay the tragic workplace accidents and fatalities which still occur, most industries in the West now operate at historically-unprecedented high standards of health and safety with an amazing delta between the records of similar industries in developing countries.

Where unions would have moderated the dangerous behaviour of industry a hundred years ago, criminal legislation now exists to harshly punish any company director found guilty of negligence resulting in death or serious injury to their workforce.

Most western economies experience fewer than 4 fatalities per 100,000 workers per year (USA has a rate of 3.3, for example) whereas developing countries such as India and China report around 10 times that number and, due to questionable data collection methods, probably experience an even worse rate in reality.

A hundred years ago, the equivalent figures for countries such as the USA, Australia and the UK would probably have been closer to the present day Chinese and Indian statistics.

Unions in the high cost economies, therefore, now focus on pay and conditions as much, if not more, than workplace safety.

As the pay rates of the workforce increase, the labour arbitrage business case of shifting operations to a lower cost economy becomes more tenable.

If an aggressive union successfully negotiates regular pay rises, especially under the threat of industrial action, which aren't in line with similar or adjacent workforces, it won't take much persuasion for the Chief Executive Officer to investigate alternatives to employing a large workforce in an expensive economy.

Train drivers and station staff on the London Underground have regularly closed the city's public transport network down through strikes, costing the London economy millions of pounds in lost productivity and disruption.

This is despite the fact "Tube" drivers are already paid very well compared to similar workers (£50,000/US$70,000 for a Tube driver versus £22,500/US$33,000 for a bus or truck driver) and have extremely generous holiday allowances (43 paid days off per year in a country where 25 is the standard).

Therefore it would be particularly surprising if there isn't already an entire team of consultants and project managers employed by the Transport for London (TfL) organisation tasked with investigating and implementing driverless train technology.

Robots don't go out on strike and don't spend eight and a half weeks on holiday each year.

TfL is a government department. Traditionally, government departments have not embraced offshoring as an option to deliver services to their customers, mainly due to politicians' fears of reprisals at the ballot box by an electorate angered by stories of local jobs being replaced by overseas workers.

This is changing, however.

Government departments have been using outsourcing as a method to deliver services for decades.

The received wisdom has long stated that it would be political suicide for a government to offshore domestic jobs, but this doesn't mean the same outcome doesn't occur by stealth.

By awarding contracts to local companies, who may deliver some of the services from overseas, government departments are able to gain the cost benefits of offshoring with an element of "plausible deniability" of the methods used to achieve those savings.

In Australia, a state government has recently decided to be a little more open and honest about this type of arrangement. The State Government of New South Wales outsourced its services department (providing human resources, finance, payroll, IT and other back office functions) to Infosys and Unisys.

In an historic move,[27] jobs were created in India in June 2015 to provide government services to residents of Sydney and the state of New South Wales.

An initial offshore department of 90 jobs was announced at the launch of the six year contracts with more to follow.

Despite significant protest from unions, opposition politicians and left-leaning commentators, the contracts were signed, the services transitioned and have been in operation for the last year without any evidence of the sky falling in as a consequence.

Even the department name is itself instructive;

**ServiceFirst** - a clear indication it is no longer acceptable to consider a government department to be primarily about local jobs for local residents but, instead, about the quality and the value of the services being provided to the ratepayers.

The initiative to outsource and offshore the ServiceFirst department is estimated to be producing a $20 million per annum saving for the ratepayers of New South Wales which, considering that prior to the change, the department was running a $17.5 million deficit, is presumably quietly welcomed by the silent majority of rate payers.

## The city that outsources everything

In the USA, the city of Sandy Springs, Georgia, has gone far

27 *http://www.itnews.com.au/news/nsw-offshores-jobs-in-215m-servicefirst-outsourcing-404715*

further; they've "outsourced everything"[28].

Sandy Springs has outsourced all government-provided services, from maintaining the parks to cleaning schools.

With the exceptions of the police and fire service, all other government departments are little more than vendor managers, monitoring pre-agreed Key Performance Indicators (metrics to measure and track quality of delivery) and issuing new invitations to tender for services which are due for renewal.

In the words of interim City Manager, Oliver Porter in 2013, "I tell every city official I meet: Your main job is not to supply jobs — it's to serve taxpayers".

## Digital Roadkill - The Consequence of Teaching in the Rear View Mirror

Hopefully it's now becoming painfully obvious the career paths we have previously followed are unlikely to remain open to our children.

Many activities which previously kept millions of people employed are simply going to disappear over the coming years either through elimination but mainly through automation and robotics.

Those functions which remain are going to be picked off as candidates for offshoring, leaving very few jobs in Western economies which the current generation of workers will recognise.

It should be abundantly clear the next generation will require expert career advice and access to targeted training to ensure

---

28  *http://www.michigancapitolconfidential.com/18713*

they don't become, in the words of the broadcaster Paul Wallbank, "digital roadkill".

What Paul means by that colourful expression is we're at risk of developing a generation totally unequipped to compete in the global economy and who stand no chance against the rise of the machines.

Of those who do choose a technology career path, if they fail to graduate in the top percentile they will find their main competition for work will be graduates from Bangalore and Shanghai.

Imagine the societal consequences of millions of young people without work or a chance of ever getting meaningful employment.

It's the point at which Friedrich Hayek's classic work on political consequences, "The Road to Serfdom" commences, inspired by his experiences and observations in pre-Second World War Germany.

The statement that by being born in a Western democracy, we have won the lottery of life, might be rapidly-decreasing in its accuracy.

What use is a cradle to grave welfare state if those who follow you can no longer be relied upon to pay into it?

If the whole premise of the generous state handouts and largesse relies on a Ponzi Scheme-like layer of new contributors being added to the bottom of the pyramid, but their prospects of longevity of employment is declining, the house of cards collapses.

Back in the Southern Hemisphere, the New Zealand Opposition

Leader, Andrew Little MP, gave a State of the Nation address in January 2016[29] which came very close to articulating the problem facing the next generation of Kiwis, referencing the example of the driverless cranes being installed at the Port of Auckland replacing a well-paid workforce.

Like most politicians, his solutions miss the point. His silver bullet is "retraining" but completely fails to mention a single skill which would be of use in the economy of the future.

We must ask ourselves, if the people we elect to make the decisions required to future-proof our livelihoods are unable to describe how they intend to do this, perhaps the problem is bigger than we previously might have thought?

Certainly, we shouldn't hold our breath waiting for the political class to play catch up and we should take ownership to find the solutions to our problems.

If our politicians haven't worked out what to do, have the schools, universities and career counsellors spotted the trend and, if so, what plans have they put in train to address the future problems?

I wish I could give you a positive answer detailing an inspiring vision and dynamic plan to completely re-focus the skills we are teaching the next generation of students and apprentices. Sadly, the opposite is more likely.

The education systems of most major Western economies are maintaining both course and speed as if we were living 20 years ago.

The USA has been churning out an average of 166,000

---

29 *https://medium.com/@nzlabour/andrew-little-s-state-of-the-nation-2016-230e91ab1323#.nd9p4pl41*

accounting graduates every year since the early 1990s and the trend is increasing.[30]

Putting aside the fact it's completely beyond my understanding why anyone would wish to embark on a career in accountancy in the first place, what on earth do the educational institutions think these people are going to do in a decade's time?

Today, I could hire an accounting graduate from an excellent university in India with three years' post graduate experience for a third of the salary of their Western equivalent.

An American or Australian graduate might be top of his or her class of 200 students but they are competing with several hundreds of thousands of graduates in India, China and the Philippines, all of whom would be happy to work for a lot less than the American or Australian could afford to accept.

USA Accounting Graduates

As illustrated in the chart above, this year in the USA over 200,000 graduates will leave university with a Bachelor's degree in accounting.

If they manage to find a job which primarily uses the skills they learned during their studies, they will be spending most

30 *https://www.aicpa.org/InterestAreas/AccountingEducation/ NewsAndPublications/DownloadableDocuments/2015-TrendsReport.pdf*

of their working day staring at an accounting programme on a computer, producing reports and interpreting the results - all of which can be performed by their Indian counterpart in Bangalore on a significantly lower salary.

The picture is the same across all manner of subjects which are unlikely to be of any use in the near future, yet still the universities keep producing graduates in IT, business finance, law, journalism, web design, and so on.

At best, these people are going to end up in a job completely unrelated to the subjects they spent three years of their life studying.

More likely, though, they'll struggle to maintain any kind of fulfilling career and will bounce from one short-lived job to another as the available positions reduce year on year.

They will begin to feel like fish stranded by the tide in a rapidly-evaporating rock pool.

Of course, these graduates will be marginally-ahead of those who are currently studying subjects which are already of dubious value in an employment environment where large swathes of job families are being eliminated, automated and offshored.

Helpfully, these degree courses all seem to have the same suffix; "Studies", as in Media Studies, Communications Studies, Social Studies, et cetera.

Douglas Adams foresaw these workers in useless careers in his Hitchhiker's Guide to the Galaxy series of books.

The residents of the planet Golgafrinchan persuaded all the hairdressers, telephone sanitisers, account executives, TV producers, public relations executives et al that the planet was under an impending threat of destruction by a mutant star goat and they should evacuate on a specially-built spacecraft, the Ark B, to colonise another planet.

In reality, the spaceship was programmed to crash land into a distant planet.

Sadly, this option is unlikely to be available to us in the near future, so we will face the possibility of park benches full of unemployed Media Studies graduates sleeping rough and bothering us for spare change.

Not only are universities failing to address the challenge of what to teach future graduates to protect their future employment prospects, they are also being slow to embrace the elimination, automation, offshore opportunities for their own businesses.

There are around 150 universities in the UK offering law degrees. Let's assume each university employs five lecturers on a salary of £55,000, a combined salary bill of £41 million.

It doesn't take a huge stretch of imagination for a university to video record an entire law degree's worth of lectures and sell them under a pay per view service.

Perhaps one of the country's most effective law lecturers is close to retirement and is prepared to have his or her last year of work recorded for posterity and to top up their pension with the viewing subscriptions?

What's the future for the country's least-effective law lecturers at this point?

The Massachusetts Institute of Technology does this already[31]

---

31 *http://ocw.mit.edu/courses/audio-video-courses/*

with a huge amount of intellectual property available freely from the University's website.

Other learning institutions have similarly made huge swathes of learning material available online to the public for free, with websites such as www.mooc-list.com providing collated lists and links to this goldmine of knowledge.

At present, it is isn't possible to graduate directly with this material but it would be the next logical step and not particularly difficult to implement.

Once this next innovation has been made, how many university lecturers are still required?

Universities have gone through several cycles during their existence; from theology-based finishing schools for the privileged few, to centres of scientific investigation for curious offspring of the independently wealthy to the present day situation where a degree, regardless of its future relevance, is seen as a necessity for any future job application.

Perhaps, if the supply of degree courses fails to match the job market of the future, we might see a contraction of courses offered or even a reduction of the number institutions offering them?

Louisiana State University recently came close to declaring bankruptcy[32]. Could this be the future for many other universities?

---

32 *http://www.nola.com/politics/index.ssf/2015/04/lsu_academic_bankruptcy.html*

## Inflexible Economy Example #1 - Australia

Australia is the land I call home and there is a special place in my heart for the self-proclaimed "lucky country".

We have indeed been exceptionally lucky in comparison to our peers in the G20.

The last official recession was in 1992 and, thanks to some good old fashioned make-work schemes such as handing out cash to citizens to buy new televisions and government grants for dubious home insulation improvements, Australia managed to navigate through the global downturn after 2008 without officially entering recession.

However, a more significant factor than the Australian government's largesse with our taxes and future earnings in the avoidance of recession was the Chinese government's decision to stimulate its economy by an infrastructure building programme of historically unprecedented scale.

More concrete was used by China in the years between 2011 and 2013 than the USA used in the entire 20th century[33].

Much of the raw material for China's building works came from holes dug in the Australian ground.

Whilst this was obviously great news for the Australian economy and any Australians employed in the extraction and export of natural resources, it rang more than a few alarm bells for those students of history who had heard of a phenomenon called Dutch disease[34].

---

33  *http://www.forbes.com/sites/niallmccarthy/2014/12/05/china-used-more-concrete-in-3-years-than-the-u-s-used-in-the-entire-20th-century-infographic/#21457f777194*
34  *https://en.wikipedia.org/wiki/Dutch_disease*

This is an observed problem which links the stimulus of a large income of foreign currency (usually from exporting natural resources such as gas or iron ore) with a long term "hollowing out" of a country's economy as all other exports become uncompetitive by comparison due to the strengthening of the currency.

Another term which may be applied to Australia is banana republic.

This might seem somewhat unfair but consider the generally-accepted definition of the expression;

A country with political instability, largely dependent on the export of a single limited-resource product.

Now consider also the fact that, between 2009 and 2016, Australia has had a new Prime Minister every year, three of whom assumed office as a result of bloodless intra-party coups with no reference to the electorate.

Reliance on a single non-manufactured export and a new national government every year? If it quacks like a duck and waddles like a duck, it might just be a duck!

Australia also has the shortest term (three years) of national government of most western democracies which might also contribute to exacerbating the issues which all high cost economies will face in a future of elimination, automation and offshoring of vast swathes of work.

If a government spends the first year of their term learning how to govern, the second year actually governing and the third year preparing to fight an election, it doesn't leave much time to build a long term strategy to deal with a world where all but a few career paths will remain available to the next generation of school leavers.

The mining and resources boom has not been immune to the relentless wave to eliminate, automate and offshore either.

To move the millions of tonnes of minerals from the mines to the new ports on the Australian coast, thousands of miles of train lines have been built.

Whilst the initial construction was labour intensive and attracted many Australians from the main suburban areas to work on a FIFO basis (Fly In, Fly Out), now the railways are in place, the trains are driverless using automation software and are monitored with GPS and track sensors.

Take note, London Underground Tube drivers, the next time the union asks you to vote for industrial action.

The massive trucks which drive the material from the base of the open cast mines to the railway loading points are, in many cases, also driverless.

Driverless trucks don't fail drug and alcohol testing and don't require two weeks off with an expensive flight home after two weeks of work in a hostile environment.

In fact, the driverless mining trucks might be a good predictor of how driverless cars could become a reality; the mining company can tightly control the movement of vehicles and pedestrians within the site and therefore the risk of unpredictable human error, which is probably the greatest threat to the viability of driverless vehicles, is negligible.

Not only do the mining companies embrace automation, they are voracious offshorers of work too.

Rio Tinto, for example, has been running back office financial activities for many years via partners such as Infosys, who employ hundreds of staff in India to service their operations.

As previously mentioned, the last major economic downturn in Australia was in 1992. In the meantime, many other western economies experienced two more with the "dotcom" and the sub-prime crises.

In one respect, Australia was lucky but, in counter-intuitive way, this may make Australia poorly-equipped for what is happening now and will continue to for the foreseeable future.

Imagine the situation where two aspiring business managers started work at the same time in the USA and Australia in the late 1980s.

By 2016 our American and Australian managers have climbed the corporate ladder to middle or top management in a major corporation but with one important difference.

The American manager will have had to have quickly learned how to cut costs and try to maintain profit margins in a difficult market whilst being responsible for a significant sized budget…. twice.

The Australian would have been too young during the last downturn to have had budget responsibility and could have only witnessed it as an observer.

I've seen this experience gap play out in several Australian companies in recent years with middle aged business managers with all the seniority and gravitas one would expect for someone at the peak of their career struggling to make clear and timely decisions in response when market conditions take a rapid turn for the worse.

Cancelling the subscriptions for the magazines in the head office foyer and having the potted plants removed isn't really going to fix things if you need to take 25 per cent out of your annual costs.

I've witnessed 50 year old executives with decades of experience struggle to grasp the concept of Pareto analysis of cost categories to determine which elements are addressable and which they have little or no discretion over.

Ultimately though, in a downturn, many organisations have only one large cost category over which they have significant control; staff wages.

When the time comes to make sweeping reductions to the cost base, the number of staff and the level at which those who remain are paid are the two main levers most large organisations can pull quickly.

Selling of assets or negotiating discounts from suppliers take time and are not guaranteed to yield results with as much certainty as firing 20 per cent of the workforce and freezing or reducing the pay of the remaining 80 per cent.

Australia has a creeping problem in this area.

The Australian industrial relations environment is globally unique in the way minimum pay rates are mandated by Federal and State governments down to a level not specified by any other nation.

An example of this is the way café waitresses and waiters are paid. In New South Wales, a government body sets and regulates the minimum hourly rates of these staff depending on their duties, setting the rates for weekday, overtime, weekends, public holidays and loadings for casual and part-time work.

This legislated rate is called the Modern Award.

In an almost Kafkaesque reference, the non-standard hourly rates are termed Penalty Rates, as if an employer is to somehow be publicly vilified for offering an employee the opportunity to increase their income.

Many other countries set a minimum wage, true, but no other G20 country sets minimum wages by industry, job type and to such a low level of detail, preferring the efficiency of the labour market to find the correct level above the low tide of the minimum wage.

The very real effect of this can be experienced by walking past a row of cafés and shops in an Australian tourist area on a public holiday or trying to find a replacement car tyre on a Sunday; many business simply choose to remain closed during these peak pay rate periods because the staff costs make their product uncompetitive.

It gets worse.

Large unionised workforces can negotiate Enterprise Agreements with employers, the minimum rates of which are already mandated by the Modern Awards, with the effect very few employees ever actually earn as low a rate as the modern award and have a far more lucrative pay scale instead.

The problem is, these Enterprise Bargaining Agreements are set for a period of several years (usually at least three); it's not hard to realise what that means for an employer in an industry which enters a significant downturn in year two of an EBA negotiated at the peak of the market.

We are seeing this play out in the resources sector now with companies who are locked into high labour costs unable to compete against those lucky enough to have set theirs during the off-cycle.

Predicting the future of an economy is fraught with opportunities to be proven very wrong especially in these post-capitalist times of huge government interventions in previously free markets and the socialising of bad debts taken on by privately-owned banks.

If Australia were to enter an extended period of negative growth, it's hard to be convinced there is enough flexibility in the structure of the workforce or experience in the executives to enable a quick or sustainable recovery.

It may in fact prove to be a burning building with locked doors as companies find their only options are to fast-track already nascent initiatives to eliminate, automate and offshore.

## Inflexible Economy Example #2 - Europe

The European Union clearly has some immediate problems to deal with, regardless of the bleak prospects for the work force of any high cost economy that we've explored already in this book.

The EU and, more specifically, the Eurozone is suffering from a seemingly never-ending currency and debt crisis, imposed "austerity" measures on the previously profligate southern states, desperate rates of youth unemployment (nearly 50 per cent in countries such as Spain, Italy and Greece), the rise of neo-Nazi groups and almost constant debt re-negotiations using threats to exit the Euro as a tactic. The result of the recent UK "Brexit" referendum has not calmed this instability either.

In addition to what seem to be "business as usual" issues in the Eurozone, there are long term issues which may pose more material problems as the drive to eliminate, automate and offshore continue. Many of the laws and regulations in place in Eurozone countries combine to result in a highly inflexible workforce.

The difficulties involved in firing employees in France are the stuff of modern folk lore. In theory it is not particularly difficult or unusual to "let go" an employee for "cause" (gross negligence, for example) or poor performance, in reality, it is practically impossible to sack an employee on a contrat à durée indéterminée (indefinite contract) with a labyrinth of bureaucracy to be navigated with multiple points of appeal for the employee all of which are heavily-weighted against the employer.

A 2013 OECD study[35] confirms that France has some of the most restrictive and inflexible job protection regulations.

In the case that an employer finds that business demands have shrunk and the workforce is too large for the available work, there are similarly complicated steps which to be completed to avoid expensive claims later.

Workers' Councils must be consulted to determine that the lay-offs are required for economic reasons. The regional regulatory body (Direction Régionale des Entreprises, de la Concurrence, de la Consommation, du Travail et de l'Emploi) must also be kept informed throughout the process to ensure compliance with the relevant legislation and, where 50 or more employees are impacted, the employer must also produce a draft job preservation plan to demonstrate that they have attempted

---

35  *http://stats.oecd.org/Index.aspx?DataSetCode=EPL_OV*

to find alternate positions for those impacted and considered filling all existing vacancies prior to the lay-offs.

What's wrong with providing such protection to the employee against exploitative bosses, we might ask? Probably not so much in an economy which is healthy and growing but in a stagnating economy an inability to shrink the workforce is a debilitating liability which is likely to reduce the chances or at least increase the time to achieve recovery for a company in financial distress.

If it's difficult to fire employees, companies are less likely to hire them in the first place.

A comparison of the Gross Domestic Product (GDP) performance of France and the USA since the global financial crisis in 2008 hints at this axiom; both countries suffered a drop in economic activity after the Lehman Brothers collapse but the USA recovered within the year and has overtaken the previous highs while France has experienced a "double dip" recession and growth has stagnated.

As Boris Johnson (Mayor of London 2008-2016) flippantly commented, "The EU is a graveyard of low growth; the only continent with lower growth is currently Antarctica".

Correlation isn't causation but the ability to shrink a workforce in response to reduced demand must certainly have been a factor in the speed of the USA recovery.

France's restrictions on labour flexibility are not unique in the Eurozone. The OECD study found other Euro countries such as Germany, Italy, The Netherlands and Portugal have even more restrictive labour regulations than France.

What might be the consequences of these restrictions in an interlinked global economy where market forces require an ever-increasing velocity of elimination, automation and offshoring for commercial organisations to simply remain profitable?

If the CEO of a multinational company were to be presented with a new business opportunity which required the quick mobilisation of a new workforce to deliver a product or service, how likely would they be to hire this workforce in France or Germany?

New sources of revenue don't come with guarantees of success and the CEO will be very aware of the possible requirement to demobilise the new workforce if the new business isn't successful. He or she is unlikely to choose a location with one of the world's most lengthy and bureaucratic processes to retrench staff if other options were available.

In a different but similar way that Australia's highly-regulated minimum wage(s) are having the unintended consequence of accelerating the destruction of jobs by locking wages into a permanently uncompetitive level, the Eurozone countries labour regulations are reducing the time for the business case for projects to eliminate, automate and offshore new work.

## What Should Governments Do?

To revisit the questions we posed at the start of this chapter;

**Is there anything governments can do to stop the destruction and exodus of jobs?**

No. There's very little other than self-defeating legislation and trade restrictions which are likely to trigger tit for tat retaliation. For every restriction placed on the provision of services with the aim of protecting the status quo, the market is likely to find a loophole or method of bypassing it.

Uber, for example, solved the problem of expensive government regulation and a cartel-like hold on the provision of private passenger transport by enabling anyone with a car and smart phone to become a private taxi driver.

Decades of governmental meddling to placate vested interests were neatly side-stepped because the delta between the true cost and the demanded price of the service had become so wide.

Ultimately though, most attempts by governments to restrict markets fail in their objective through either incompetence or corruption.

In the words of American pundit Cal Thomas;

"One of the reasons people hate politics is that truth is rarely a politician's objective. Election and power are."

Replace "truth" with the words "long term benefits to the citizens" and the quote still holds true.

### Even if governments could stop it, should they and what are the alternative courses of action?

There's a sense that, even if there were a method to put a brake on or reverse the trend of the elimination and automation of work or its movement to low cost economies, it would just be delaying the inevitable.

A country which manages to halt these technological advances or reduce the input cost of goods and services is likely to be storing a far worse set of issues for the future, not least of which is increased inflation, as the rest of the world adapts to cope with the new reality.

Perhaps it would be better to accept the trend is a reality and will be for some time to come (perhaps forever) and expend effort on working out what the future economy is likely be composed of.

Depending on the direction of your political leaning, you might feel your government should be planning more effectively for this new reality or perhaps just getting out of the way of the free market doing its job.

Certainly there is a role for the education sector in ensuring training for redundant skills are phased out and replaced by those which will be of value for the coming decades. There are some "soft skills" and attitudes which will be of use too and we will discuss these in a later chapter.

Anyway, the jobs have already gone! In many industries and sectors, as Phil Fersht of Horses for Sources explains, the horse has already bolted (excuse the pun) and the foundations for expanding the scope of elimination, automation and offshoring have already been laid and simply need to be built upon further[36].

---

36 *http://www.horsesforsources.com/jobs-wont-be-jobs-anymore_022016*

# Chapter 6
# Work categories still in danger

If you were trying to select a future-proof job, what activities would you need to avoid?

Perhaps the simplest way to investigate this is to look at the targets people like me select when we are asked to remove complexity and cost from an organisation.

The first category is probably irrelevant to a future job-seeker; presumably nobody knowingly chooses a career where the entire purpose of the job has been long-forgotten and people are doing that way "because that's how we've always done it".

However, that's clearly what must have been going on in the British Leyland vehicle factory in the 1970s.

My targets for elimination, automation and offshoring tend to fall into the following broad categories;

- Any task which involves a large proportion of the time on the phone or sat in front of a computer.

- Any task which doesn't require high quality interaction with clients.

- Any task which isn't revenue-generating.

- Repetitive activities with non-complex work rules.

- Any activity with a large delta between local rates and overseas rates.

• Militant, inflexible or unreliable workforces.

A 2013 study examined the probability of jobs which currently exist today being automated within the next 20 years. The results might come as a surprise[37];

The Likelihood Of A Profession Being Automated

| Profession | Percentage |
| --- | --- |
| TELEMARKETERS | 99% |
| TAX PREPARERS | 99% |
| INSURANCE APPRAISERS | 98% |
| SPORTS UMPIRES AND REFEREES | 98% |
| PROCUREMENT CLERKS | 98% |
| BOOKKEEPING, ACCOUNTING & AUDIT | 98% |
| REAL ESTATE BROKERS | 97% |
| SHORT ORDER COOKS | 94% |
| TAXI DRIVERS | 89% |
| HIGHWAY MAINTENANCE WORKERS | 87% |
| SUBWAY AND STREETCAR OPERATORS | 86% |
| PARKING ENFORCEMENT OFFICERS | 84% |
| HOME APPLIANCE REPAIRERS | 72% |
| COMMERCIAL | 55% |
| INTERPRETERS & TRANSLATORS | 38% |
| ACTORS | 37% |
| FLIGHT ATTENDANTS | 35% |
| FIREFIGHTERS | 17% |
| ELECTRICIANS | 15% |
| AIR TRAFFIC CONTROLLERS | 11% |
| REPORTERS | 11% |
| TRAVEL AGENTS | 10% |
| PHOTOGRAPHERS | 2% |
| FASHION DESIGNERS | 2% |
| MARRIAGE & FAMILY THERAPISTS | 2% |

The study agrees with our suggestion in the chapter "Automation", that there is a high probability (55 per cent) the job of a commercial airline pilot may no longer exist.

While we might feel a little uncomfortable with the idea the

37 *http://www.oxfordmartin.ox.ac.uk/downloads/academic/The_Future_of_ Employment.pdf*

plane we're about to board has nobody at the controls, perhaps we can all raise a celebratory glass at the 97 per cent probability that real estate agents will no longer exist.

Who knows, perhaps there might be an opportunity for an aspiring citizen journalist blogger with a smartphone to catch an iconic photograph for posterity recording the very last real estate agent buying their final cheap suit from the soon to be redundant last department store shop assistant?

There are some surprises thrown up by the study, though; is there really a 98 per cent chance of sports umpires and referees or a 37 per cent chance of actors' jobs being automated?

Are these such radical suggestions though?

Think about how technology is already being used to judge difficult "line/ball" decisions in tennis or the way the action replay is used to check whether a try was scored in rugby matches.

Consider also how many on-screen actors appeared in the film Avatar.

The epic battle scenes in the Lord of The Rings films were almost entirely produced by CGI (computer generated imagery).

Maybe automation in these areas are not such a ridiculous idea after all.

# Chapter 7
# Which jobs might remain?

Making predictions is a fool's game and hours of amusement can be had by simply looking back at hugely inaccurate newspaper articles, TV documentaries and movies from the recent past which tried to predict how we might live in the future.

When he was transported to 2016 in the film Back to the Future, Marty McFly found he could ride a hoverboard but still had to use a phone booth to make a call (albeit, a video call).

As the Frey and Osborne study referenced earlier admits, predictions about which jobs might disappear or remain can be nothing more than exercises in probability.

Looking at the current and increasing trend towards elimination, automation and offshoring, we can see some common factors in the work which is hardest to remove or reduce.

There are several themes and aspects of work which suggest it will remain reasonably intact for future generations.

In my view, I will always struggle to remove jobs which have the following attributes;

- You have to physically be "in the room"
- You must personally consult with individuals
- You have a deep understanding of a local "culture"

- You produce artistic or creative work or prestige products
- Where the key to success is in the quality of "the relationship"

If we wanted to further distil these five attributes down to a single statement, it might be the following;

*There is only a future for careers in expensive economies where the value or quality of local knowledge or presence outweighs the cost advantage of every other method of performing the task.*

We'll now explore some possible aspects of career paths which might survive a little longer in the Western economies.

## You have to be physically "in the room"

This is likely to be the employment route the majority of the future generations will find themselves heading down.

The volume of jobs in this category will be tradespeople, the skilled workers who build and fix things which can't easily be imported.

For as long as there are significant controls on immigration, there will always be a premium paid for construction workers, plumbers, electricians, painters, plasterers, gardeners, and so on.

While these roles might be impacted by some level of elimination through the use of imported "sealed spares", (as in the case of car repairs) and automation, it's possible to foresee there will need to be human oversight for some significant time to come, simply due to the reasonably unique nature of each individual job.

We could build a machine to install the power circuits, sockets and lighting in a standardised design of a newly-built house or apartment but it would require a far greater effort to ensure the same machine could retro-fit a 30 year old house without ripping up the plaster and floorboards by mistake.

The move towards "sealed spares" in the car repair industry is an interesting example of how work which remains onshore may become "dumbed down".

A car mechanic 25 years ago may have been expected to remove, dismantle and repair components such as an alternator or carburettor, whereas his or her modern day equivalent simply swaps out the faulty part with a brand new piece.

In general, most "in the room" jobs that will remain are likely to be blue-collar, manual work and over time may trend towards becoming less skilled; the "eyes and hands" to install complex components at site rather than designing or repairing those complex components.

However at the top end of town, surgeons and dentists will be difficult to automate and, despite some routine medical procedures already being performed "down the wire", it's probably going to be a very long time before any of us choose to go under the surgeon's knife for complex surgery when we are in New York and the scalpel is held by a robot controlled over the internet by a surgeon in Manila.

The greatest flaw in the logic of the predictions of medical procedures being carried out remotely is that there isn't a guaranteed service level for the Internet. If the connection is not available or congested with other traffic, you can't sue anyone for breach of contract for a "buffering timeout".

Many routine medical services might be delivered remotely however.

We may get used to the idea of seeking health advice and prescriptions from a General Practitioner whom we speak with over a video call or we might seek a repeat prescription from an automated service. Indeed, the New Zealand Government is trialling this in some rural areas.

There's some regulatory hurdles to overcome to deliver this service, the doctors will need to be registered and approved to practice medicine in the country they are delivering services to, but these might be overcome quite quickly once a business case has been established.

## You must personally consult with individuals

Management consultants are the modern day equivalent of snake oil salesmen; we use our charm, references and qualifications to convince clients our exorbitant day rate is worth it.

In fact, the higher the day rate, the less likely it is the client will want to doubt the credibility of the advice they are receiving.

Very few consulting engagements are paid for on a risk/reward basis where the consulting firm will guarantee the business outcome the client requires. We have been expecting clients to wise up and demand measurable outcomes from their expensive annual expenditure on the small army of consultants for years now.

While the day rate model remains the norm, our longevity is

assured though.

Other professions which have a significant enough component of consulting for some of the roles to survive the impact of elimination, automation and offshoring include advertising, architecture, medicine and sales for high cost, quality or prestige goods and services.

The changes to the work environment, the high levels of unemployment and temporary nature of many of the remaining jobs might cause some psychological issues for a large proportion of society.

Consulting with an Indian psychiatrist over Skype about your psychological pain from losing your job to offshoring might add insult to injury, so there may well be a solid career path with healthy prospects of longevity open for professions such as psychiatrists, therapists, addiction counsellors, and others.

## A deep understanding of a local "culture"

As much as we might loathe the wall to wall advertising we are inflicted with on a daily and hourly basis, there is a very definite line between effective marketing and terrible failures.

Knowing the market you are trying to sell to is critical. This is something that is probably impossible to replicate by an algorithm and is highly problematic to replicate from outside of that culture.

Even today, badly overdubbed foreign TV adverts are cheap but not particularly effective compared with highly-targeted locally-written versions.

Offshore call centres expend significant effort to improve the knowledge and linguistic skills of their customer service agents in an attempt to close the gap between the two cultures.

It's far from effective though, and a call centre agent in Bangalore wishing a customer in New York, "have a nice day" or a Sydneysider, "g'day" is still a long way from being culturally aligned.

Organisations which might otherwise have cut costs by offshoring client-facing staff such as those answering the phone when you call your bank have realised this and many now make a positive marketing point out of the fact their customer service centres are in the local country.

Of course, this doesn't prevent the majority of back office functions being provided elsewhere.

Taking this a step further, the UK bank, First Direct, realised Britons associate a higher level of trust to people with Scottish accents and therefore relocated some of the client contact centre departments to Scotland. There is likely to always be a market for such services which are provided at a slight premium to the consumer with the guarantee they will speak with someone within their national borders.

## Artistic or creative work and prestige products

Humankind has always craved the beautiful and the rare. This trait is unlikely to disappear from our psyche any time soon. If you can find a niche providing things of beauty to a paying customer base, you'll have a career with longevity.

This is traditionally fraught with pitfalls. Most famous artists lived a hand-to-mouth existence during their lifetimes, and very few artists ever see the wealth their efforts later generate.

Lower down the artistic scale, home designers, personal stylists, hairdressers, and others should still be with us performing reasonably similar roles to those of today.

## The key to success is in the quality of "the relationship"

It's hard to imagine the sales representative of a company which has a high value, complex product being particularly successful if they were not located physically close to the key decision makers in their target customer base.

Google, Microsoft or whatever company is the next Google or Microsoft are going to be unable to make multi-million dollar sales to large corporations using phone calls and video conferences alone.

Some level of corporate schmoozing, wining and dining will always be necessary.

Therefore, if you've got the gift of the gab and can close big deals, there might still be a career path for you.

There is a problem though; how do you move from small deals to large deals if the entry level jobs no longer exist for you to develop those deal-closing skills?

## Is there an alternative option?

Further exploring the problems caused by the removal of "entry level" jobs, perhaps there's an unforeseen consequence in the fact that these first roles of so many professions will eventually only exist in overseas locations?

What will this mean when we are recruiting the senior and executive versions of the profession in later years?

If, for example, it is determined by the Board that a Chief Financial Officer needs to have an in-depth understanding of accounting principles, yet for the previous two decades the majority of the intermediate positions within organisations have been based out of countries such as India or the Philippines, how will a company find a suitable candidate?

There might be only three choices;

1. Select a candidate without the traditional accounting background, or

2. Select an experienced candidate from one of the overseas companies which have been providing these accounting skills, or

3. Have an exchange programme, where those graduates deemed to be of high potential spend a year or two working in the overseas delivery centres learning the trade, grooming them as future CFOs.

There are pros and cons to all three options; the first option would require a significant change in traditional thinking on what a CFO needs to know to perform his or her role.

Option two would need some lobbying of government immigration departments to ensure that work visas could be secured.

Option three would be the most significant development to indicate a coming to terms with the inevitable.

It would be an acknowledgement there are simply some careers which are no longer performed in our Western economies.

## What might the future workforce look like?

The majority of present day job types in the categories described above as likely to remain are "blue collar" trades.

Currently, these are filled by men, either because historically these trades were seen as suitable only for males or the roles were not seen as attractive careers for women.

In a work environment where there is far more competition for fewer jobs, perhaps this might change as women begin to compete with men to be plumbers, plasterers, electricians, and so on.

The pool will have shrunk but there will still be the same number of fish competing for food.

It has already begun, with trades' training organisations actively encouraging women to move into these traditionally "blokey" roles.

The remaining work categories (local knowledge, artistic, relationship, consulting, et cetera) already have more diverse workforces and this trend may well continue.

# Chapter 8
# Skills your children will need

## Time management

The days of working for a single employer for your entire career are long gone for most of us already.

This will remain the case for future generations as work will come from multiple sources and clients/employers (and the distinction between the two will become blurred).

A worker in what Paul Wallbank and others have termed the "gig economy"[38] will need to balance clients, work locations, new channels for work (such as social media, smart phones, et cetera) and their home life.

This isn't so radical an idea though; to refer back to the generational changes touched upon in the introduction, for a large period of human history, most people worked in a self-employment model to some degree. Large corporations and companies are a recent phenomenon and, in the Anglosphere's Common Law, these entities first had a legal standing from the 16th Century.

"Clocking on" at 9am and finishing work at 5pm may still be a possibility for some workers but, increasingly, the lines between work time and home life will blur. Maintaining the balance will be challenging and many of us will struggle to get it right

---

38  *http://paulwallbank.com/2016/01/09/working-in-the-gig-economy/*

whilst ensuring a healthy pipeline of revenue in a world where we are constantly bombarded with electronic reminders of our workload.

In addition to the competition between work and home life, the new freelance workers in the gig economy will need to maintain and improve their skills and qualifications and have an eye on the future to ensure that their profession isn't about to become the next victim of elimination, automation and offshoring.

In highly-competitive markets, procrastination tends to be punished quickly and severely.

There are well-established tools and techniques to teach people effective time management and teachers of these methodologies are likely to find themselves increasingly in demand.

Good time management is a core skill of any successful businessperson, but in an environment where increasing numbers of people work for themselves and for multiple clients, the difference between success and failure, or indeed, happiness and frustration, will be an ability to juggle all competing demands on the limited resource of time.

## How to structure your finances

For those entering the brave new world of freelance working, the days of fixed amount pay cheques each month are over.

What you earned last year will not be a reliable indicator of what you might earn this year and your finances will need to be structured accordingly.

Domestic and business expenses will need to be capable of

being flexed to reflect unpredictable reductions in demand and income; any costs which can be paid as a unit charge with the flexibility of being switched off and on at short notice will be more attractive than long term commitments.

This may pose a challenge to lending institutions as mortgage payments might be made one month but not another, requiring flexibility of lending terms.

Ironically, as the entry level accounting roles are being automated or offshored, the remaining workers in the high cost economies will need a basic understanding of accounting principles, or at least enough to ensure that they are on top of their cash flow and have provisioned enough to cover the end of year tax bill.

As today's freelance workers already know, choosing the correct legal vehicle to work within and from which one receives income is critical; get this wrong and nasty tax surprises can ensue or worse.

## How to structure your attitude

In an economy and work environment where previously stable and solid jobs are rapidly disappearing or being dumbed down, those who manage to remain gainfully employed or start their own businesses will need a very different attitude to work and to receiving income than the generations before them.

New realities will have to be faced by workers in the gig economy.

Related to the requirement to be able to effectively manage time and competing demands, gig workers will need to consider that work might arrive at any time of the day or night, especially if the services they perform are not bounded by international borders or time zones.

If so, do you want to accept the work and if not, will you lose out because someone else has taken it from you?

Also consider your competitor doesn't necessarily live in the same city, country or time zone as you and their costs might be significantly lower than yours.

If so, what is the edge you have over them? Your Unique Selling Point (USP) will need to have at least one of the aspects that we explored in the previous chapter, such as that you need to be "in the room", you're consulting with the client, you share a deep understanding of a culture, you're offering artistic or prestige products or the goods and services rely on a close relationship with the client.

Working smartly is going to be a critical consideration for those who will be successful. We will all have to become experts in continuous improvement.

A small business owner or freelance worker needs to ask themselves, "Of the work I do, which activities take up the most time?"

Take the cost of that time and multiply it by a thousand; would it equal the development costs to invent a different way of doing it?

How about when multiplied by a million? This is how software automation and offshore service providers look for opportunities.

If what you do looks a likely candidate, either try to invent the solution yourself, sell the concept to someone who can invent it or switch careers before they do.

On a similar note, the gig economy will have the same risks as the job categories which have already disappeared to elimination, automation and offshoring.

Smart workers in this new reality will need to constantly be aware that what made money this year probably won't make as much next year and might not make any money the year after.

The gig workers will need to always have an eye over their shoulder looking out for disrupting ideas and inventions.

One way of checking the risk level is to ask, "If I were entering this industry as a know-nothing kid, what are some of the idiotic questions I might ask? What would need to change for these questions to suddenly not seem so stupid?" The large taxi companies which are now seeing Uber strip away their revenue might have benefited from the answers to these questions if they had asked them several years ago.

## Personal branding and social media

Before the Internet, teenagers could hold embarrassing opinions and make poor decisions with a low risk of there being lasting negative consequences.

This is no longer true in a world where nothing we post on the Internet is biodegradable.

Deleted pictures and posts on social media are recoverable by those so minded to search for them.

Traceable unprofessional content posted on the Internet may

come back to bite the originator and "retweeter" worse than an embarrassing choice of tattoo design or location of body piercing.

If in any doubt and you feel like you absolutely have to express yourself for the world to view, use an anonymous account.

Conversely, the Internet (and social media in particular) could be the biggest and best shop front to your personal brand.

The larger your network in the real world or online, the greater the chance of an unexpected opportunity finding its way to you.

Until you've found a niche, it may be useful to keep your LinkedIn or similar Internet-based business network profiles quite generic to reduce the risk of missing opportunities by being pigeonholed.

## Self-sufficiency and beware of other people

If fewer of us are going to be working for big organisations in the future, we are unlikely to receive the benefits and protection provided by a large human resources department.

Workplace bullying isn't going to be handled in a consistent way in multiple small companies as it might have been in a company of several thousand employees.

Worse, if many of the co-workers are now subcontractors or freelancers, there will be even less motivation for the HR team to get involved as they won't be viewed as full employees.

In reality, HR departments have never really been there for the protection of the workforce, but for the protection of the productivity of the workforce.

A 2011 study[39] found that one in 20 business leaders may be sociopaths, i.e. unable to feel much, or indeed, any empathy for others.

In many careers such as politics and management, this lack of humanity is actually a distinct professional advantage.

Dick Fuld, for example, managed to survive multiple decades without ever falling foul of the HR departments of the various companies he bullied and manipulated his way through to become the final, disastrous CEO of Lehman Brothers.

So, in a situation where we're all freelancers and the HR department is going to be less motivated to help and, as we've seen in an earlier chapter, neither is the government, self-sufficiency and a thick skin are going to be important positive character traits. Actually, they've always been useful life skills but their importance is going to have a resurgence.

In her 2006 book, The Sociopath Next Door, Martha Stout describes the signals which indicate you're dealing with a sociopath and, more importantly, how to deal with one (spoiler; the only way to deal with a true sociopath is to run away).

The four per cent of the population who have potential sociopathic tendencies aside, working in the gig economy comes with other people-based risks and identifying these and taking appropriate mitigating actions will be a core skill required for the gig workers.

Assessing the likelihood of being paid on time, in full or even at all could be the difference between solvency and bankruptcy

---

39 *https://www.theguardian.com/science/2011/sep/01/psychopath-workplace-jobs-study*

for someone working on an hourly rate for just a few clients.

Even if all of these skills are in place, it may be you've selected to work in the next profession to be eliminated, automated or offshored.

In this case, quickly identifying this new reality will be critical otherwise you'll risk becoming your profession's equivalent of that last video rental store on the high street, watching the revenue decrease month on month but somehow paralysed, unable to take the inevitable final step of closing the shutters and finding something else to do.

How will you know this is about to happen to you? When a smartly-dressed consultant arrives at your place of work and is introduced as the "Continuous Improvement Manager" or "Project Manager" for a project you've not previously heard of.

# Chapter 9
# What to do if you've already chosen a doomed career

So far, we've looked at jobs and entire careers which have fallen victim to the relentless and inevitable drive to eliminate, automate and offshore work from high cost economies and we've explored aspects of those few professions which might remain.

But what should you do if you have invested much of your adult life to a career choice which is under threat in the future from the same disruption airline navigators, accounts clerks and garbage collectors have already experienced?

Sadly, your choices are limited and mainly not very palatable.

## Denial isn't a river in Egypt

This is clearly not going to be strategy with a happy ending, but it is the one most people will choose.

It will work out for a few people; even in industries which are declining there will be a final few who have worked there until the bitter end.

Most will have seen their work and income dry up long before, however.

History has proven this isn't a strategy which will work for the majority, from the 19th Century textile workers in the Northwest of England who smashed power looms and spinning machines (later to be known as "Luddites") to the many disruptive strikes held by the London dockers throughout the 1960s and early 1970s as their industry was transformed and no longer required their services.

The modern equivalent of this spirit of denial is the rush to petition government to introduce further regulations and restrictions.

Most unions seem to have chosen this special pleading as their preferred tactic despite there being plenty of lessons from the recent past indicating the unlikely long-term benefit of this.

In the west, we have more government regulation of industries and professions than ever before.

These regulations produce opportunities for "entrepreneurs" to profit from artificially-closed markets and government-created monopolies.

## Match the Indians' pay rates

Again, this is unlikely to have a happy ending but, for jobs most at risk of offshoring, the only real way to defuse the threat is to reduce the delta between the overseas workers and the high cost economy workers pay.

Typically, the labour arbitrage for white collar, office-bound work between, say, the UK and India is about 30-40 per cent after project and management overheads are included.

If this were closer to 10 per cent, the project cost and additional management overhead would convince most corporations the benefit of offshoring is not worth the effort and disruption.

Of course, it will be completely impractical for most to take an arbitrary cut to their take-home pay of at least 50 per cent; bills still need to be paid and the high cost economy will hold no sympathy for someone who is having to compete with and match the cost of a worker in a low cost country.

The tough psychological impact of this choice is also going to be difficult to overcome for the most workers.

## Relocate/Emigrate

The difference between high and low cost economies is, of course, a sliding scale. Currently, Australia tops the charts for wages relative to anywhere else.

If your skill set and career is tied to an industry currently under threat in Australia, perhaps moving to a lower cost economy might be the answer, at least for a while?

Of course, the logical end to this approach is that, if you want to keep doing the same work until retirement, you might ultimately have to move to India or China and work for the local wage.

There may always be a domestic relocation option though, as those organisations which have made a unique selling point (USP) out of their "local service" will always require in-country staff to answer phones in their contact centres.

If your skills and experience are of value to these organisations, moving to the area where these services are provided from might still be a viable option, if you're very lucky they may even run a "virtual" call centre allowing you to work from home.

The career prospects for anyone in this domestic niche will be increasingly limited though.

## Retrain/Reskill

This is fairly obvious; if you're working in a doomed industry or profession, you could choose an alternate career path, retrain and start again.

Depending on how much time, effort and emotion you've invested in the job which is about to be made redundant, this could be one of the biggest life choices you will have to take.

Choosing a new career which is complementary to and benefits from the skills and experience of the previous one might mitigate the disruption, but, depending on your personal financial situation, you will probably suffer a decline in your standard of living and may struggle to maintain important commitments such as mortgage or rent payments.

This is likely to be the outcome for the vast majority of workers under threat.

## Be the disruptor

This is the choice with the biggest potential return on investment. If you can predict your career is a likely target for disruption from elimination, automation and offshoring, why not put serious effort into thinking how you can personally benefit from this before and during the disruption?

Not everyone has the entrepreneurial spirit or access to capital required to be the next Malcolm McLean or Travis Kalanick but somebody else always will.

If you don't have the funding or the spark of the idea to start another industry's version of Uber, perhaps you can start something on a small scale that is still large enough to pay the bills whilst removing the risk of being made redundant by the inevitable changes approaching.

Don't look to this author for suggestions though; if I knew, I'd be building the business now rather than writing books about it.

**The End**

# About the Author

Rob Gaunt advises organisations on how to benefit from the opportunities available through elimination, automation and offshoring of work.

He lives in Sydney, Australia and can be contacted via his website – www.eliminateautomateoffshore.com or his LinkedIn profile - https://au.linkedin.com/in/robertgaunt

## Acknowledgements

Most of this book was written by the swimming pool during two family holidays, so thanks and apologies must go to Renate and our kids for my anti-social behaviour and lack of attention to the various diving tricks and acrobatics on display. Hopefully the ice-creams went some way towards compensating the younger family members.

I'm indebted to friends and colleagues who have contributed to the ideas and have found rogue apostrophes and grammatical errors especially Stuart Hall, David Watterson, Katie James and Paddy Lewis.

Thanks also to Tony Khoury of The Waste Contractors and Recyclers Association of NSW for his briefing on the history of the developments in domestic waste collection.